Lecture Notes in Networks and Systems

Volume 805

The series "Lecture Notes in Networks and Systems" publishes the latest developments in Networks and Systems—quickly, informally and with high quality. Original research reported in proceedings and post-proceedings represents the core of LNNS.

Volumes published in LNNS embrace all aspects and subfields of, as well as new challenges in, Networks and Systems.

The series contains proceedings and edited volumes in systems and networks, spanning the areas of Cyber-Physical Systems, Autonomous Systems, Sensor Networks, Control Systems, Energy Systems, Automotive Systems, Biological Systems, Vehicular Networking and Connected Vehicles, Aerospace Systems, Automation, Manufacturing, Smart Grids, Nonlinear Systems, Power Systems, Robotics, Social Systems, Economic Systems and other. Of particular value to both the contributors and the readership are the short publication timeframe and the world-wide distribution and exposure which enable both a wide and rapid dissemination of research output.

The series covers the theory, applications, and perspectives on the state of the art and future developments relevant to systems and networks, decision making, control, complex processes and related areas, as embedded in the fields of interdisciplinary and applied sciences, engineering, computer science, physics, economics, social, and life sciences, as well as the paradigms and methodologies behind them.

Indexed by SCOPUS, INSPEC, WTI Frankfurt eG, zbMATH, SCImago.

All books published in the series are submitted for consideration in Web of Science.

For proposals from Asia please contact Aninda Bose (aninda.bose@springer.com).

Andrea Bencsik · Anastasia Kulachinskaya
Editors

Digital Transformation: What is the Company of Today?

 Springer

Editors
Andrea Bencsik
Department of Management
University of Pannonia
Veszprém, Hungary

Anastasia Kulachinskaya
Graduate School of Industrial Economics
Peter the Great St. Petersburg Polytechnic
University
St. Petersburg, Russia

ISSN 2367-3370 ISSN 2367-3389 (electronic)
Lecture Notes in Networks and Systems
ISBN 978-3-031-46593-2 ISBN 978-3-031-46594-9 (eBook)
https://doi.org/10.1007/978-3-031-46594-9

This Springer imprint is published by the registered company Springer Nature Switzerland AG
The registered company address is: Gewerbestrasse 11, 6330 Cham, Switzerland

Paper in this product is recyclable.

Contents

Transformation of Marketing Complex of IT-Companies in the Digital Age (The Case of Video Game Industry)

Nikolay Nikolaevich Molchanov⊙, Oksana Sergeevna Muraveva⊙, Kirill Antonovich Yumashev⊙, and Nikolay Vladimirovich Lukashov⊙

Abstract The digital transformation of industries and companies is an integral element of progress at the present stage. However, companies are experiencing problems in systematizing and finding effective marketing digital approaches. The purpose of this paper is to develop proposals for improving the marketing mix for digital video game industry companies. Interest in the study is due to the lack of effective marketing tools that would satisfy the requirements of modern companies in this field. The methodological basis of the paper is the research of theory and practice of applying innovative digital technologies in the field of marketing. An empirical study of video game companies was also carried out on the base of general scientific principles, methods of systemic, logical analysis, generalizations, as well as methods of mathematical statistics. The results of the research can be summarized in three blocks. Firstly, an overview of the digitalization level of countries. For this, the relationship between the overall level of digitalization and the level of GDP per capita by countries was analyzed; business use of the main digital technologies was investigated; index of digital technologies development was determined by country of the world. Secondly, the characteristics of the video game industry were given: the concept, market capacity, main segments, the place of the Russian Federation in this market was determined; the numerical value of companies in the video game industry was analyzed for different types of markets; ranking of countries by the size of the company's net profit in the video game industry was presented. Thirdly, the peculiarities of marketing mixes elements for IT products were indicated, with the highlighting of new components suggested by authors; the financial performance of companies using streaming platforms as a technology to promote their products were analyzed; the main digital marketing technologies were identified in terms of popularity among IT companies. A comparative analysis of specific marketing software was also carried out, which can be used in the development of marketing campaigns in the IT field to create, analyze and optimize the reach of the target audience.

N. N. Molchanov · O. S. Muraveva (✉) · K. A. Yumashev · N. V. Lukashov
Department of Business Economics, Entrepreneurship and Innovation, St. Petersburg State University, Universitetskaya nab, 7–9, 199034 St. Petersburg, Russian Federation
e-mail: o.muraveva@spbu.ru

Keywords Digital transformation · Video game industry · Marketing mix ·
Innovation · IT-companies

1 Introduction

Many sectors of the economy are currently undergoing digital transformation
processes [1]. With the changing paradigms of the consumer and business market,
organizations are increasingly forced to adopt digital technologies. This is already
an unstoppable process, and digitalization is a concomitant element of progress. The
continuous development of Information and Communication Technologies (ICT) has
led to the emergence of a new digital reality, where new sectors, products and services
have been developed as a result of the rapid digitalization of the world economy [2].

When forming the Global Innovation Index (GII) 2022, experts identified the main
directions for two new waves of innovation: (1) a wave of innovation in the field
of digitalization based on supercomputers and artificial intelligence (ICT-sphere)
(2) a wave of Deep Science innovation based on breakthroughs in biotechnology,
nanotechnologies and new materials. According to statistics, Russia is 47th in the
GII ranking among 132 countries. Today our country is in 34th position in terms of
the use of ICT, and only in 70th in terms of the export of ICT services (1.7% of the
total turnover) [3].

At the present time companies from the digital industries are experiencing issues
in systematizing and finding effective methods for products promotion, because
of the emergence and development of new technologies. Innovations in the technology sector affect the process of digitalization in industries, in digital marketing
itself and in digital products. The development of digital marketing, and the decline
in the effectiveness of classical marketing tools, as well as the increase in prices
for such campaigns, requires the systematization of theoretical provisions and the
development of a methodology for using digital technologies in marketing.

A distinctive feature of a digital product from a classic physical product is its
intangible form. This entails both certain difficulties (legal problems of authorship
and patenting of such products, pricing issues, and others), as well as additional
opportunities. The most important and basic opportunity is the ability to innovate
and upgrade the product endlessly. At the same time, digital product can be updated
and upgraded without any technical and time restrictions. They also contain the
fundamental developments of computer science and mathematics, in particular Data-
Science and Computer-Science [4]. The growth of digital sectors of the economy not
only depends on R&D, but also forces to strengthen fundamental and applied research
for ever-increasing technological needs. The highest priority in such areas are the
achievement of quantum superiority, the creation of virtual reality and devices that
allow human consciousness to interact directly with the virtual digital world [5].

In the last two decades, digital marketing has been actively developed and studied,
using digital technologies (for example, PCs, mobile phones, the internet and its
elements, etc.) to promote products and services, research target markets, sales,

provision of services, community management, etc. Its progress has changed the way companies use marketing tools. As digital technology advances at an accelerating pace and costumers use digital devices more and more, digital marketing campaigns are becoming predominant and actively developing now.

Modern marketing concepts are less developed in the outdated classical markets, but are actively being used in digital industries with digital products. But even companies in the video game industry that went digital more than 15 years ago face significant challenges in finding and applying new digital technologies that meet their requirements in marketing campaigns. A few years ago, every fourth product of the industry could boast of good profitability, but in modern realities, due to the increased importance of marketing and the obsolescence of its classic tools, as well as the poor research on this issue, only every twentieth product is recognized as commercially successful [6].

Many authors deal with marketing issues using digital technologies, in particular Gavrilov [7], Kulikova [8], Amado [9], Faruk et al. [10], Butenko and Chernikov [11], Grigoriev and Chvyakin [12], Krasavina [13], etc.

There are much fewer works that would highlight the problems of digital sectors of the economy and innovative digital products presented at them. Such problems were considered in the papers of the following authors: Druklna [14], Kalimullin [15], Sokolova [16].

There are very few studies devoted to the marketing of videogaming digital products. Interest in this field is due to the lack of effective marketing tools that would meet the requirements of modern companies in the videogaming digital markets.

2 Material and Methods

The theoretical and methodological basis of the paper was the research of leading authors involved in the theory and practice of applying innovative digital technologies in the field of marketing. The study used reference and information publications, factual information published by marketing and information agencies of statistics, information materials and articles, as well as legislative acts and individual laws of different countries.

An empirical study (questionnaire) of companies in the video game industry, the results of which are also presented in the article, was carried out based on general scientific principles of a comprehensive study of economic phenomena, methods of systemic, logical analysis, generalizations, as well as methods of mathematical statistics. Selective and monographic observations were used. The questionnaire designed by the authors of the study was distributed via the internet. The collected data were processed using IBM SPSS Statistics 26 & Excel program. The following tools were used to analyze the data: frequency analysis, comparison of means, correlation analysis.

3 Results

3.1 Overview of the Overall Level of Digitalization of Countries

There is a pattern that countries with high GDP per capita have a higher digitalization index than countries with low GDP per capita (Table 1) [17, 18]. However, Russia breaks out of this logic of interconnection. It has an above-average digital transformation index, with a low GDP per capita. This can be explained by the availability of digital technologies (the cost of the internet, services, etc.), low tax rates, preferential terms for digital companies, etc. Digital methods of production and consumption of the product are transition factors to digitalization. The software, music and video game industries are the pioneers that have almost completed the digital transformation.

According to Rosstat statistic, at the end of 2021, the share of organizations with a website was 46.2%. If we compare this value with EU countries, we can see that 90% of large EU businesses have their own websites [19].

The situation with the use of traditional ICT by Russian companies [20] is the following. The share of enterprises using ERP for doing business by medium and large businesses reaches 19.6 and 37.5%, respectively. If we talk about the EU countries, these values reached 28% for small businesses, 57% for medium businesses and 76% (for large businesses). So, for example, in Belgium for medium-sized companies, the value reached 77.5% [21].

The share of companies using CRM-systems in Russia reached 13.4%. The maximum values among the OECD countries for this indicator were: Germany (43.35%), the Netherlands (62.18%), Finland (77.89%) respectively [22].

Table 1 The level of digitalization of countries, thousand USD per year (2021)

Country	Region	GDP per capita, *thousand USD*	Digitalization, %
Singapore	APAC	73	97
USA	North America	70	88
Sweden	EU	61	86
Israel	East	52	75
Germany	EU	51	95
South Korea	APAC	35	100
Russia	BRICS	12	55
China	APAC	12	94
Brazil	BRICS	7,5	39
India	APAC	2,2	52

Table 2 Use of key digital technologies by businesses in Russia and the EU

Digital technology	In Russia (%), 2021	In EU (%), 2021
ERP	13,8	38
CRM	13,4	35
SCM	4,8	34
Digital platforms	14,7	78

The share of Russian enterprises using SCM systems, reaches 4.8%, while 41.26% in Germany and 66% in Belgium. The share of Russian enterprises using digital platforms is 14.7%, but in EU this figure reached 78% [23].

The summarized result is presented in Table 2, which demonstrates statistics on the main digital technologies and averages for Russia and the EU.

Therefore, the level of use of digital technologies by Russian business is much less than that in the EU, while the maximum gap between the EU countries and Russia is in SCM technology. This conclusion can be confirmed on the base of the Information Technology Development Index (ICTD Index), which for Russia is 7.07, which is above the average, but below the advanced countries of the EU and the USA. So, Fig. 1 shows that Iceland has the maximum index (8.98), and the United States in this rating takes 16th place in the global list [24].

If we look at the digital industry, Russia occupies only 0.6% of the entire global digital industry. For the video game industry, which is part of the digital industry,

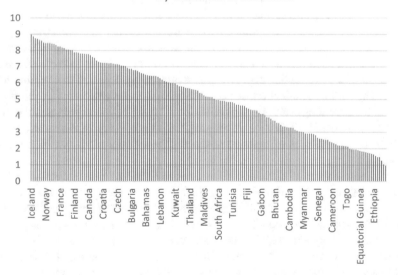

Fig. 1 Index of development of digital technologies by countries of the world. (Compiled by the authors)

this value reaches 1%. According to Russoft statistics, the size of the digital market in Russia reached $34 billion and grew by 8% compared to the previous year [25].

Among specific digital technologies in the Russian online space, Yandex, the main search engine, dominates, followed by Google. Among the most popular applications are Yandex Taxi, Yandex Money, Yandex Market and others. Yandex is mainly used in Russian, takes into account the culture and its shades. SEO in Russia also has its own particular qualities.

Digital transformation actively promotes the implementation of digital technologies, including in the marketing of IT-products. For example, there is a change in the structure of regional sales, by simplifying the globalization of product sales. It does not make sense to limit sales to a particular region or country in the video game market, since distributors allow products to be sold online in all countries where this is possible.

3.2   Peculiarities of the Video Game Market

A video game is a digital product which is sold online digitally and doesn't have a physical body or representation. The video game industry covers many different work disciplines such as programming, design, animation, modeling, etc. The manufacturing process in the industry is one of the most complex in the global economy, requiring the involvement of world-class specialists to solve problems such as real-time physics simulation, material strength calculation, simulation of the behavior of light and the reflective properties of the surfaces of material objects, etc. The industry employs tens of thousands of people around the world.

The high growth rate of the industry affects other industries, such as information technology, software, computer components (hardware), microelectronics, space industry, and others. The development of the industry and the growth of the market affect the entire economy as a whole, which in turn contributes to the development and growth of the production capacity of personal computers and their components, contributing to the introduction of new technological developments in the field of information technology and in the production of computer components, which are used in most industries of the world economy, from the creation of films and computer graphics, to calculations of manned space flights and the most complex physical simulations.

Since the creation of each new item does not require raw materials and does not incur any costs, then, therefore, marginal costs will always tend to zero, unlike ordinary physical products. As well as marginal cost, the cost per unit of output will tend to zero–decrease in proportion to volume. Only 1–5% of market products can be called commercially successful, it depends on the segment, but earlier this number exceeded 25%. Because of this, enterprises tend to develop several of their products simultaneously to minimize risk, and if they fail, they may exit the market. In order to make a profit and cover the budget in this case, the product often needs to pay off ten times or even more.

According to the annual report of the marketing agency 'WCP', the market has more than one and a half billion unique consumers, where 20% of the market are console users (300 million users), and 67% of the market are personal computer users (1 billion users). Also, more than 80% (1 billion 200 million users) prefer mobile platforms such as smartphones, set-top boxes, etc. [26]. Among the markets, according to this study, the largest one is the Asian market segment, which occupies 33% of the total world market (500 million people). The European market ranks second and occupies 20% of the global market with 300 million consumers, both on personal computer and mobile platforms. At the same time the market for console video games in Europe and North America is more than twice the size of Asia and reaches 100 million people.

The age of a typical customer is around 34 years. The number of men is 6% higher than the number of women and equals 56%. The typical gamer spends an average of 7 h a week playing multiplayer (network) games. Among online game players, 36% prefer shooter, 28% prefer action, and 27% prefer adventure [27].

The video game market and industry are divided into three main segments: AAA, AA and Indie. Each of them has its own consumer and its own competition model. These segments are divided according to several indicators:

1. Number of employees. However, it is important to take into account the peculiarities of the digital industry. There is a similar situation with gradations of business sizes. For example, according to a business size assessment, Valve, having a staff of 150 people, would be considered a medium or small business, however, based on net profit calculations from 1 billion USD to 10 billion USD per year will be considered big business.
2. Project budgets. It is generally accepted that an Indie project cannot cost more than 100 thousand USD, while an AA does not exceed 10 million USD and AAA projects can cost more than half a billion USD [28].

The largest AAA-players (EA, Activision-Blizzard, Ubisoft, Saber-Interactive, Nintendo and others) occupy more than 95% of the market. The rest of the market share is occupied by low-budget AA and Indie studios. Based on the growth of capitalization and profits of these companies, one can judge the profitability and prospects of the market as a whole. In general, it is customary to consider the AAA segment of the market to be oligopolistic, with a small number of large players, while companies in the Indie and AA segments compete freely with each other. At the same time, companies from lower segments can move to higher-budget segments. This is explained by the fact that if the product shoots and sales volumes increase many times, then due to the specifics of production in the industry, the costs do not increase, and this allows companies to reinvest a significant share of the funds.

The authors analyzed the numerical value of video game industry companies in countries with a developed financial market, in countries with emerging and frontier financial markets. For this stage of the study, 1803 video game companies from 54 countries were selected from open data banks, and their quantitative share by country was determined (Table 3).

Table 3 Quantitative distribution of video game industry companies by countries

№	Country	Number of companies	(%)	№	Country	Number of companies	(%)
1	United States	662	37	10	Poland	23	1,3
2	Japan	379	21	11	Australia	22	1,2
3	United Kingdom	248	14	12	Czech Republic	22	1,2
4	Canada	59	3,3	13	Netherlands	20	1,1
5	Germany	53	2,9	14	Russia	20	1,1
6	France	52	2,9	15	Finland	16	0,9
7	Sweden	38	2,1	16	Spain	14	0,8
8	South Korea	31	1,7	17	Taiwan	10	0,6
9	China	25	1,4	18	Other countries	84	4,66

It was revealed that majority of companies in the video game industry are in the US, Japan and the UK (37%, 21% and 14% respectively). These countries are followed by Canada, Germany, France and Sweden (3%-2%). All other countries, including **Russia** (1,1%), is less than 20% of total companies.

If we look at the spread of companies by country in terms of developed, emerging and frontier financial systems, then 94% of all companies are located in countries with developed financial markets, 5%–in countries with emerging financial markets (including Russia), and the remaining 1%–in frontier financial systems. The correlation coefficient between the degree of development of the financial market and the number of companies on it is **0.881**. It is quite obvious that the more developed the financial market, the greater number of companies from the video game industry is there.

Further, the profitability of video games companies was assessed among the countries (Table 4) [29]. The main major markets of the world are represented: China, USA, Japan, Great Britain, South Korea, Germany, France, Canada, Italy and Spain. Russia is not included, as its net income of the video game industry is below 2 billion USD.

The level of net profit in the video game market in China reaches 40 billion USD, which is 4 billion USD more than in the United States. This is due to the peculiarities of the economic development of the China and the growth in the number of its population–the attitude towards video games in the Asian region differs from the entire main world, the size of the China video game market is much larger than the video game markets represented in other countries. It is also important to note that India with the same population, is a much less digitalized country than China, which serves as evidence in favor of the theory about the peculiarity of the Chinese video game market.

Table 4 Ranking of countries by net profit of video game industry companies, billion USD per year

Country	Net profit, billion USD per year
China	40,85
USA	36,92
Japan	18,68
The Republic of Korea	6,56
Germany	5,97
Great Britain	5,51
France	3,99
Canada	3,05
Italy	2,66
Spain	2,66

As a result of data analysis, it becomes obvious that the digitalization level and development of the video game digital industry in countries with developed financial markets will be higher than in countries with emerging and frontier financial markets. The results obtained can be confirmed by the assumption that in countries with developed financial markets, credit is cheaper due to greater competition among banks and, in general, due to their larger number, which means a more favorable environment for doing business. Secondly, in such countries there are more necessary specialists (digital products development requires qualified personnel) due to a higher level of education.

3.3 The Use of Digital Technologies in the Marketing Complex for Video Game Companies

According to current research, global digital marketing spending reached 450 billion USD in 2022 and continues to rise, estimated to reach 500 billion USD by 2024. Digital marketing spending in Russia reached 6 billion USD in 2022 [30]. The Russian market is unique, mainly characterized by its local platforms such as Yandex, or the leading social network VK with a market share of 83%, which owns platforms such as Mail.ru, etc. The Russian Federation has one of the fastest growing digital markets in the world: Russian e-commerce (electronic commerce, microtransactions, etc.) in 2020 reached revenues of 33 billion USD, also stimulated by the SARS COVID-19 pandemic [31]. By 2024, it is expected that total size of Russian e-commerce markets will reach 300 billion USD. At the moment, according to Rosstat, the share of e-commerce in the entire GDP of Russia reaches 3.29%. Among the top positions of countries with the greatest demand for digital marketing are such countries as Canada, India, USA, UAE, Australia, Ireland, Philippines, England.

When considering specific digital marketing technologies that are practiced by Russian IT-companies, the following can be distinguished:

1. O2O marketing–includes Big Data analysis and the subsequent use of predictive analytics, as well as the effective interaction of offline and online tools. The market for such services has reached 30 billion rubles [32]. It includes the development of personalized offers, alternative cheaper methods of delivering information to the desired segment, the ability to collect more data on consumer behavior, evaluating the effectiveness of online advertising, etc. Many digital companies describe the experience of testing such marketing as positive (for example, VK, и Odnoklassniki reported revenues increased by 15% and 4% accordingly [33]).

2. SEO (Search Engine Optimization)–increasing the visibility of the site or information in the search engine, in order to stimulate the transition to a media resource through keywords in the search. Each search service uses its own SEO algorithms, such as Google Ads, Bing, Yandex.Direct, and others. SEO, from one point of view, can be considered its own media. SEO also has an earned media component where search engine visibility can be improved by getting relevant backlinks from websites. According to the experience of Russian digital companies (for example Gameloft) that published the effectiveness of this digital marketing tool, SEO attracts almost 10 times more traffic than regular social networks. This tool is actively used by IT-companies in Russia and around the world.

3. CRO (Conversion Rate Optimization)–conversion rate optimization (capturing traffic from paid advertising and increasing its conversion), used to increase conversions on the site, is a complex digital marketing tool. The end result of this tool is to increase in the profitability of an advertising campaign. This tool is rarely used compared to others, but some digital companies (for example Mail.ru) practice this digital tool.

4. SMM (Social Media Marketing)–includes complex interaction with the target audience on a specific digital platform or social network. Used by almost any IT-company.

5. Digital PR (Public Relations)–increases brand awareness and works with the media and the target audience in the media space (in social networks, blogs, podcasts and other digital resources). Digital PR also includes reaction to negative or positive brand mentions online and public relations through a website, such as a social media news hub or blog.

6. DP (Digital Partnerships)–is creating of partnerships for mutual benefit with various digital platforms, advertisers or public figures on digital platforms. This digital tool is actively used by IT-business, especially by video game companies in the world, and in Russia in particular. For example, the Russian company BattleState Games, using only this digital tool, increased sales of its product by more than 10 times [34].

In general, the practice of foreign companies in the use of digital technologies in marketing is the same as the practice of Russian companies. However, as statistical data show, Russian companies use digital technologies less frequently than Western companies.

3.4   Study of Effective Digital Marketing Tools Based on a Survey of Representatives of IT-Companies

An empirical study was carried out by the authors of the article to explore the experience of video game companies in the framework of their marketing campaigns. The questionnaire survey was offered to a number of Russian and foreign companies for which financial data for the latest reporting periods was available. As a result, the most popular and effective digital marketing tools among video game providers were identified. To confirm the conclusions, the effectiveness of the identified tools was compared with the financial performance of the companies.

In total, 22 companies from different countries, including Russia, took part in the survey. Both large companies and small "Indie-studios" were covered by the survey.

Authors planned to find out: the role of marketing in the video game market over the past 10 years; degree of agreement with the statement that a well-designed marketing campaign is currently a decisive factor in the success of a product in the market; evaluation of the contribution of community management to the success of the project; which of the digital marketing technologies is the most effective now in the video game market; what marketing technologies will be mostly in demand in the future; estimation of the percentage of financially successful products in the video game market.

The following results were obtained:

1. All of respondents (100%) believe that over the past 10 years, the role of marketing in the financial success of digital products in the video game market has grown significantly. The role of marketing still plays a paramount role and the percentage of financially unsuccessful products continues to grow. (As part of the 2015 study, one of the results was the calculation of the proportion of financially successful projects, where the criterion for financial success was taken as a net profit in the amount of 10 times the project development budget [33]). As part of this current study, an attempt was made to correlate these data with current data to date. As a result of data processing, a result of 7% was obtained. An increase in the share of financially successful projects by 2% was revealed. One can try to explain this by the subjective judgment of the respondents about projects within the framework of their companies' experience.

2. Respondents consider the marketing campaign to be the decisive factor in the financial success of the project. It was revealed that the vast majority of respondents (~85%) consider marketing to be the fundamental element in the financial success of the video game market product. But it is inefficient to promote video game market products using classical promotion methods.

3. Currently, community management plays one of the most important roles in the success of a product. Due to the peculiarities of consumer behavior and close contact between the target audiences of projects and developers and publishers, one can observe very intensive community management on the part of companies. As a result of the analysis of the data obtained, the following conclusions

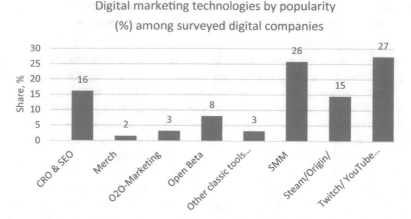

Fig. 2 Digital marketing technologies by popularity (%) among surveyed digital companies

can be drawn: some (30%) of the respondents believe that community management is extremely important in the market, the majority, in turn (40%), consider
community management important, but a well-constructed marketing campaign
and the quality of the final product are more important.

4. The respondents were given a list of the following main digital marketing
   technologies: "Twitch/Youtube (Digital Partnership)", "SMM", "Open Beta",
   "Internal tools of digital distribution sites", "O2O-Marketing", "CRO and CEO".
   Options were also offered: "Classic methods of promotion" and open answer.
   On Fig. 2 is a summary of the popularity of digital marketing tools used by IT-
   companies responding to the video game industry. The y-axis shows the share of
   references to digital technologies in the practices of marketing campaigns (%).
   The abscissa shows specific digital technologies provided by the Internet survey.

   The analysis of the received data showed that among all the technologies
   mentioned, the respondent companies turn to streaming platforms most often.
   This can be explained by the spontaneous growth of such sites on the part of
   the target audiences of various projects, as well as the significant cheapness of
   this technology compared to other digital marketing technologies. SMM and
   SEO as more universal digital marketing technologies are in second and third
   places in terms of popularity, respectively. Thus, the hypothesis is confirmed.
   Also, it is worth noting that their technologies were also mentioned, which were
   not provided for in the answer and are not digital marketing technologies (for
   example Promotional Merch).

5. Respondents disagreed about which technology will be most in demand in the
   future. They suggested that streaming platforms and influencers will remain the
   most sought-after marketing technologies in the video game market. The companies also suggested that SMM will soon be more in demand. Among the answer
   options were neuro-technologies and neuromarketing. This can be explained by
   the fact that the respondent is one of the few companies in the global industry that

is actively developing in the field of neurotechnology and in the coming years will be ready to introduce a neuro-implant that allows the user to be directly aware of himself in virtual reality.

To confirm the effectiveness of this technology, it is possible to compare the effectiveness of the MSE (marketing and sales expenses) of the part of the respondent companies that use this technology of streaming platforms and companies that do not use this it (Table 5). It is worth noting that since the authors did not have data exclusively on MSE, the SGA indicator was used, which includes: management costs, salaries of marketing service personnel, depreciation; materials and resources for the normal functioning of the marketing service and, accordingly, the costs of developing a product, price, communication, marketing and distribution policy [35].

The average SGA cost growth rate for companies was 52.85%, while the average cost-to-revenue ratio was 6.17% for companies practicing Digital Partnership tech nology and 13.79% for companies not using this technology. Thus, it can be said that companies using Twitch/Youtube and other streaming services as a digital marketing tool, on average, have more than two times lower marketing costs with proportional revenue indicators. Speaking about individual revenue growth for companies, on average it ranged from 3 to 161% for companies using the studied technology in their marketing campaigns, and from 1 to 4% for companies that do not practice it. The average annual revenue growth rate for companies has reached 40%.

Thus, several conclusions can be made: companies that practice the streaming platform technology have higher revenue growth rates, lower marketing costs growth rates, and a significantly lower ratio of marketing costs to revenue. Also, marketing costs continue to grow steadily on average in the market. We can say that the

Table 5 Revenue and SGA of companies using streaming platforms for promotion

Company	Revenue, mln USD	Year						
	2015	2016	2017	2018	2019	2020	2021	
Blizzard	4664	6608	7017	7500	6489	8086	8803	
Ubisoft	1463,753	1393,997	1459,874	1731,894	1845,522	1594,831	2223,8	
Saber	25,2496	35,3164	59,4866	475,0903	548,1346	1018,623	1657,058	
Take Two	1082,938	1413,698	1655,55	1595,182	2408,064	3187,582	3552,597	
ND	75,111,31	67,574,21	70,333,54	77,116,54	78,166,33	75,980,84	84,892	
	SGA, mln USD	Year						
Company	2015	2016	2017	2018	2019	2020	2021	
Blizzard	734	1210	1378	1062	926	1064	1025	
Ubisoft	283,572	303,633	314,416	337,087	408,292	376,374	430,6	
Saber	13,3903	20,8058	27,8232	387,9258	389,0376	390,1494	624,773	
Take Two	235,341	198,309	285,453	256,092	391,4	458,424	444,985	
ND	4063,201	3262,335	3365,455	3674,47	3477,292	3306,574	2453,252	

marketing mix is more effective with this technology than the marketing mix without the use of this technology within the video game market. Companies in the video game industry are not inclined to spend money on the development of new innovative marketing tools and prefer to use existing ones. As a result of data analysis, it was revealed that about 80% of respondents do not experiment with the development of new tools.

4 Discussion

4.1 Modification of the Marketing Complex for IT-Products (The Case of Video Games)

When developing a marketing campaign for both an IT-product and for video games in general, more parameters should be taken into account than when developing a classic 4P marketing mix. In this regard, within the framework of this study, it is proposed to modify the existing classic marketing mix into a 6P mix applicable to digital products and video games in particular.

Place as a separate element of the marketing mix complex does not have such paramount importance as for ordinary physical products. This assertion can be supported by the arguments:

1. Some IT-products assume global distribution either through individual global and local distributors, or by the company itself. Artificial limitation of sales by region exists, but in reality, is never observed for economic reasons (usually political, legal and company philosophy).
2. There are exclusively local IT-products. Marketing of such products is carried out within the scope of the distributed region. However, it is important to note that anyone in the global world market can download or purchase such a product.
3. Due to the nature of IT-products, it currently does not make economic sense for companies to have offices or retail outlets.

Because of the peculiarities of IT-products, as part of the modification of the marketing mix, two additional "Ps" were proposed–"Platform" and "Publisher". This modification realizes itself as much as possible in the video game market, but can be used by any IT-product. Each element should be considered separately.

Platform –can be considered as the device through which user interacts with the end product. This device stores an IT-product in its memory and, if necessary, provides the user with access to it. The importance of considering the platform as a separate element of the marketing mix proposed by the authors of the study lies in the fact that the platform is what often transforms the product.

Companies, should decide for which platform (or for a number of platforms) their product is being developed. This can be explained by several reasons:

1. The developing cost for different platforms is different, companies for each specific product and for each specific platform need to analyze and predict costs and correlate them with potential profit.
2. Each platform often has its own audience, which sometimes overlaps (for example, sometimes the same consumer can buy the same product both on "Play Station" and on "PC").
3. Each platform is different. Often, development for different platforms requires different specialists.

With platforms, as sometimes with distributors, the concept of "exclusivity" is often associated. For example, video game companies rarely limit product development to a single platform. However, platform owners (such as Sony with "Play Station" or Microsoft with "Xbox") can offer lucrative contracts to developers and publishers to develop exclusively for their platform, which can attract new consumers and force them to purchase the platform.

Publisher–a company that helps development companies in various aspects of business activities. Such companies exist in many digital markets, including the markets for computer software, video games, cinema, and others.

Finding a publisher is extremely important for companies-developers. The publisher can provide them with the necessary resources, including financial investments, as well as assistance in bringing and distributing products to the market. The publisher can help with marketing, distribution, and product development. For small companies or independent developers, finding the right publisher can help to move to higher segments, such as from Indie to AA. In this case, the publisher will provide financial support, which can improve the quality of the final product by expanding the staff and competencies of employees through the hiring of more qualified personnel. In reality, publishers rarely work with low-budget projects and small local companies.

Under certain conditions, the publishing company can dispose of the development companies, organizing various projects between them. Companies are to report on the funds spent, development progress, are required to adjust the product development process in accordance with the standards of the publishing company and comply with corporate policy.

Developer companies can function successfully without a publisher, implementing and distributing the project on their own. For small "Indie" companies, it is enough to get by with electronic distribution, for example, "Steam", but then the ability to enter other platforms will be limited or greatly complicated. In this case, companies should not count on significant profits, with rare exceptions in the form of "hitting" in trends. It can also be noted that companies without a publisher, if such companies are planning a marketing campaign, need an additional department, and the cost of conducting a campaign may exceed the cost of developing a product.

If there is a publisher, he assumes all obligations for the implementation of the marketing campaign, public relations and, if any, additional funding. Thus, it is possible to compare the proposed author's marketing mix with the classic 4P and 7P (Table 6).

Table 6 Application of marketing mix elements for IT products

Element of marketing mix	Classic physical product	IT-product
Product	Applicable, plays a primary role	Applicable, plays a primary role
Price	Applicable, plays a primary role	Applicable, plays a primary role
Place	Applicable, plays a primary role	Not of paramount importance for a digital product
Promotion	Applicable, as well as for a digital product	Applicable, as well as for a physical product
Process	Applicable, as well as for a digital product	Applicable, as well as for a physical product
People	Applicable, as well as for a digital product	Applicable, as well as for a physical product
Physical Evidencee	Applicable	Not Applicable
Platform	Not Applicable	Applicable, plays a primary role
Publisher	Not Applicable	Applicable, plays a primary role

Platform and Publisher are not applicable to classic physical products. However, individual firms may perform similar roles to the publisher. Authors analyzed specific marketing software that can be used in the development of marketing campaigns in the digital field. To develop any marketing campaign, various software tools designed for marketers are used.

Marketing process management software is a tool that allows companies to manage their marketing campaigns. Typically, such software includes features such as various social media audience tools, detailed analytical tools and financial reporting, content management, and others. Such software is used to create, analyze and optimize campaigns in order to maximize the reach of the desired target audience. Comparative analysis of the characteristics of these platforms has been conducted (Table 7).

The main software platforms that are used in the video game market were highlighted:

1. Marketo [36]–this platform offers automated marketing tools such as SMM, analytics and financial marketing reporting tools, etc.
2. HubSpot [37]–offers a set of tools to automate the marketing process, including online content management, SEO, CRO and others
3. Salesforce Pardot [38]–this platform is designed for B2B marketing and provides tools for lead scoring, SMM, various marketing campaign promotion metrics.
4. Act-On [39]–platform offers an integrated set of marketing tools including analytics, landing pages and more.
5. SharpSpring [40]–this comprehensive platform offers tools for marketing process automation, analytics and reporting, SMM and more.

Table 7 Comparative analysis of software platforms

Parameter	Platforms/Software				
	Marketo	HubSpot	Salesforce Pardot	Act-On	SharpSpring
Price	Starting from $1000 per month	Free and paid subscriptions from $50 per month	Starting from $1250 per year	Starting from $900 per month	Starting from $400 per month
Pricing model	Mix of models from tiers/ footprints/ adaptive use	Mix of models from tiers/ footprints	Mix of models from tiers/ footprints/ perclient	Perclient model	Tiers model
Assortment of tools	Over 40 + tools including SMM, Event and Webinar Marketing	SMM, content management, CRO, CEO, WEB analytics and more	SMM, analytics and more	SMM, analytics and more	SMM, CRO, analytics and more
Complexity	Average	Low	High	Low	Average
Regional restrictions	Yes	Yes	Significant, EU and US only	No	Unknown
Markets	Any, more aimed at classic physical markets	Aimed at IT companies, can be applied to any market	Not specialized	Not specialized	Specializing in service markets
Type	Marketing Process Automation Software	Marketing Process Automation Software	Marketing Process Automation Software	Marketing Process Automation Software	Marketing Process Automation Software
Owner company	Adobe	HubSpot	Saleforce	ActOn	Constant Contact
The target audience	B2B	B2B and B2C	B2B	B2B and B2C	B2B

Knowing the advantages and disadvantages of platforms, it is possible for companies to choose the best option that meets all the requirements. So, for Indie companies, price will be a critical factor and they should consider platforms operating in b2c markets. At the same time, the complexity of using the software will also be an important factor. Thus, for small Indie companies, HubSpot can be recommended, as the software can be used not only by corporate clients, but also by individuals. HubSpot licensing costs are significantly lower than for similar products from other companies. The focus on customers from digital markets can be considered an additional argument in favor of choosing this marketing software.

5 Conclusion

The overall level of digitalization of countries generally correlates with the level of GDP per capita. Russia breaks this logic and has an above-average digital transformation index. However, the level of use of digital technologies by businesses is significantly lower than in the EU.

The video game industry in the Russian Federation occupies about 1% of the global value. The market itself can be divided into 3 segments: AAA; AA, Indie, each of them has its own consumer and competition model.

Most companies in the industry are located in countries with a developed financial market, the top three include countries such as the US, Japan and the UK. In general, the level of digitalization and development of the video game IT industry in countries with a developed financial market is higher than in countries with emerging and border financial markets.

In the course of an empirical study of representatives of IT companies, it was found that the vast majority of respondents consider marketing to be a fundamental element in the financial success of a video game market product, and community management plays one of the most important roles in the success of a product. Most often, companies turn to streaming platforms, which can be explained by the spontaneous growth of sites on the part of the target audience, and the significant cheapness of this technology. SMM and SEO are in 2nd and 3rd place in terms of popularity, respectively. Companies that use streaming platforms as a video game promotion technology have higher revenue growth rates and a higher marketing return on investment (ROMI). It can also be said that the marketing mix is more effective with this technology than without it.

In view of the peculiarities of IT products, it would be reasonable to use two additional elements of P–Publisher & Platform as part of the modification of the marketing mix. Companies, during the development of a product concept, need to decide for which platform their product is being developed. Related to this is the notion of publisher choice and exclusivity. A publisher can assist video game producers with a number of functions. Among them: financial support; marketing and distribution; public relations and proprietary platform companies; quality control; consultation and assistance in development; development organization.

A comparative analysis of specific marketing software was also carried out, which can be used in the development of marketing campaigns in the IT field to create, analyze and optimize the coverage of the target audience.

References

1. Vaska S, Massaro M, Bagarotto EM, Dal Mas F (2021) The digital transformation of business model innovation: a structured literature review. Front Psychol. https://doi.org/10.3389/fpsyg.2020.539363
2. Petropoulos G (2022) The ICT revolution and the future of innovation and productivity. Massachusetts Institute of Technology, Stanford University and Bruegel, Global Innovation Index 2022. https://www.wipo.int/edocs/pubdocs/en/wipo-pub-2000-2022-expert-contribution3-en-the-ict-revolution-and-the-future-of-innovation-and-productivity-global-innovation-index-2022-15th-edition.pdf
3. Global Innovation Index (2022) What is the future of innovation-driven growth? Soumitra Dutta, Bruno Lanvin, Lorena Rivera León and Sacha Wunsch-Vincent https://www.wipo.int/global_innovation_index/en/2022. Accessed 08 Apr 2023
4. Smith FJ (2006) Data science as an academic discipline. Data Sci J 5(19):163—164. https://doi.org/10.2481/dsj.5.163
5. Papageorgiou A (2013) Measures of quantum computing speedup. Phys Rev A J 88(2):22–31. https://doi.org/10.1103/PhysRevA.88.022316
6. Minimum Sustainable Success. Blog edition «Gamasutra». https://www.gamasutra.com/blogs/DanielCook/20150415/241145/Minimum_Sustainable_Success.php. Accessed 23 Oct 2022
7. Gavrilov LP (2023) E-commerce: textbook and workshop for universities, 5th ed., revised. and additional. Yurayt Publishing House, Moscow, pp 563. (LP Gavrilov)
8. Kulikova ES (2020) Digital marketing: bibliographic overview. Moscow Econ J 10.400—494. https://doi.org/10.24411/2413-046X-2020-10692
9. Amado A, Cortez P, Rita P, Moro S (2018) Research trends on big data in marketing: a text mining and topic modeling-based literature analysis. Eur Res Manag Bus Econ 24(1):1–7
10. Faruk M., Rahman M, Hasan S (2021) How digital marketing evolved over time: a bibliometric analysis on Scopus database. Heliyon 7. https://doi.org/10.1016/j.heliyon.2021.e08603
11. Butenko ED, Chernikov IS (2020) Infrastructure of the digital economy: digital marketing. Bulletin of the North Caucasian Federal University, No 4(79), pp. 23–37
12. Grigoriev NY, Chvyakin VA (2019) Brand promotion in social networks by means of digital marketing. Economics 9(6A):192–201. (yesterday, today, tomorrow)
13. Krasavina V (2019) Current trends in the IT services market. In: E3S Web of conferences 135, 04039 ITESE-2019. https://doi.org/10.1051/e3sconf/201913504039
14. Drokina KV (2016) Market of information and communication technologies and organization of sales. Part II: study guide. Publishing House of the Southern Federal University, Taganrog, pp 76
15. Kalimullin K (2018) Question: how to determine the price of an IT product. https://vc.ru/ask/44450-vopros-kak-opredelit-cenu-it-produkta
16. Sokolova ES (2019) Peculiarities of the marketing of IT Companies. Beneficium 2(31):47–56. https://doi.org/10.34680/BENEFICIUM.2019.2(31)
17. Worldbank. https://data.worldbank.org/indicator/NY.GDP.PCAP.CD?locale=ru&locations=. Accessed 123 Apr 2023
18. Statista. https://www.statista.com/statistics/1042743/worldwide-digital-competitiveness-rankings-by-country. Accessed 12 Apr 2023
19. Abdrakhmanova GI, Vasilkovsky SA, Vishnevsky KO, et al (2023) Digital economy: 2023: a brief statistical collection. National Research University "Higher School of Economics". M.: NRU HSE, 2023, pp 120. https://rosstat.gov.ru/statistics/infocommunity. Accessed 07 May 2023
20. Reaping the benefits of ICT: Europe's productivity challenge. The Economist Intelligence Unit, http://graphics.eiu.com/files/ad_pdfs/microsoft_final.pdf. Accessed 23 Apr 2023
21. Digital Economy and Society. Eurostat. http://appsso.eurostat.ec.europa.eu/nui/submitViewTableAction.do. Accessed 23 Apr 2023
22. Eurostat. https://ec.europa.eu/eurostat/statistics-explained/index.php?title=E-business_integration. Accessed 07 May 2023

23. Rosstat. https://rosstat.gov.ru/statistics/infocommunity. Accessed 23 Apr 2023
24. Index of development of information and communication technologies. International Telecommunication Union. https://gtmarket.ru/ratings/ict-development-index. Accessed 23 Apr 2023
25. The state of the IT industry in Russia. RUSSOFT. https://russoft.org. Accessed 23 Apr 2023
26. Woodside Capital Partners. Marketing Agency. http://www.woodsidecap.com. Accessed 07 Apr 2023
27. ESA–Entertainment Software Rating Board. http://www.theesa.com. Accessed 07 Apr 2023
28. Deciphering Indie, AAA, and AA Games. Gameopedia. 2022-02-22. https://www.gameopedia.com/indie-aaa-aa-games-comparison. Accessed 12 Feb 2023
29. Statista Gaming monetization-Statistics & Facts. Statista. https://www.statista.com/topics/3436/gaming-monetization. Accessed 12 Apr 2023
30. Digital Ad Spendings. Marketing Agency eMarketer. https://www.eMarketer.com. Accessed 20 Apr 2023
31. Digital Marketing in Russia. Marketing Agency Giulio Gargiullo. https://www.giuliogargiullo.com/ecommerce-russia. Accessed 22 Apr 2023
32. O2O marketing in Russia. https://segmento.ru. Accessed 24 Apr 2023
33. Cook D (2015) Minimum sustainable success. 15 Apr 2015. https://www.gamedeveloper.com/business/minimum-sustainable-success#close-modal. Accessed 03 Apr 2023
34. Gelishkhanov IZ, Yudina TN, Babkin AV (2018) Digital platforms in the economy: essence, models, development trends. St. Petersbg State Polytechn Univ J Econ Sci 11(6), 22–36. https://doi.org/10.18721/JE.11602
35. Roger JB (2018) Marketing-based management. Pearson International Edition
36. Adobe Marketo Homepage. https://business.adobe.com/. Accessed 21 Mar 2023
37. HubSpot Homepage. https://www.hubspot.com. Accessed 21 Mar 2023
38. Salesforce Pardot Homepage. https://www.salesforce.com. Accessed 21 Mar 2023
39. Act-On Homepage. https://act-on.com. Accessed 21 Mar 2023
40. SharpSpring Homepage. https://sharpspring.com. Accessed 21 Mar 2023

Hierarchical Cybernetic Model of Oil Production Enterprise with Distributed Decision-Making Centers

Daria E. Fedyaevskaya⊙, **Zhanna V. Burlutskaya**⊙, **Aleksei M. Gintciak**⊙, and **Saurav Dixit**⊙

Abstract The chapter addresses the features of decision-making in multi-level hierarchical systems. The features of such systems are the set of optimum points for various individual system's components and the subsequent difficulty in coordinating and choosing optimal or quasi-optimal solutions related to conflicts of interest at different levels of management. The study describes the game-theoretic formalization of management, including control actions performed by agents, feedback flows, agents' objective functions, i.a. utility functions. The presented formalization is based on the concept of transition to the management of multi-objective hierarchical systems via multi-agent simulation models. Based on the provided formalization, the chapter describes the organizational system of an oil production enterprise, which is a typical three-level enterprise divided into decision-making levels: strategic, tactical, and operational. By analyzing the system, we developed a conceptual management model at an oil production enterprise using a multi-agent approach. Within the developed system, agents are represented as decision-making organizations: parent organization, subsidiary, administrative and managerial personnel of the field. Thus, the result of the work is a conceptual model of a three-level multi-objective system. The prospects for the research development are: detailed elaboration of the structure of the system and the decision-making tuple; development of algorithms for the multi-agent system intelligent module, which will optimize the games in the system.

Keywords Information systems · Digital transformation · Cybernetic model · Decision support systems · Multi-level objective system · Hierarchical organizational system

D. E. Fedyaevskaya · Z. V. Burlutskaya (✉) · A. M. Gintciak
Peter the Great St. Petersburg Polytechnic University, 195251 Saint-Petersburg, Russia
e-mail: zhanna.burlutskaya@spbpu.com

S. Dixit
Khalifa University of Science and Technology, 127788 Abu Dhabi, United Arab Emirates

1   Introduction

Most modern management systems are multilevel hierarchical organizational systems [1, 2]. In the simplest or highly abstracted systems, management is represented by a manager agent (supervisor) and a subordinate agent who directly interacts with the organizational system processes. The greatest interest in the field of hierarchical systems is directed to multi-level objective systems [2, 3]. In such systems, decision-making is distributed between levels. The problem of decision-making in them lies in the fact that although there is a common goal-setting, contradictions may arise both between the actions of the same-level agents and the goal-setting of agents of different levels (boss-subordinate) [2]. Researchers pay the greatest attention to the mathematical formalization of hierarchical systems, where decision-making is divided into two levels, which are considered the basic ones [4]. However, in such systems, the lower level makes decisions regarding the impact on the process of the organizational system itself.

Due to the presence of several agents with different goal-setting in multi-level systems, conflicts of interest arise, including situations in which decisions made by different agents mutually reduce their effectiveness [5]. Moreover, experience-based management leads to the adoption of not only suboptimal, but even quasi-optimal plans. To make more effective decisions, systems offering analytical justification are required. Multi-agent intelligent systems currently being developed allow modeling groups of agents with different goal-setting. This type of simulation models is a priority for use in hierarchical multi-objective systems due to their features.

In this chapter, we propose to consider a three-level target system in which decisions are distributed among agents in accordance with their classification: strategic, tactical, and operational. At the same time, modeling the basic structure of a multi-objective system will facilitate a changeover to more complex systems that are currently developing widely. Within the study, we carried out a game-theoretic formalization of the management of a typical multi-objective hierarchical system. The chapter proposes a changeover to the management of multi-objective hierarchical systems via the use of multi-agent simulation systems. The research results are illustrated with a multi-level management system at an oil production enterprise.

2   Materials and Methods

2.1   Hierarchical System Theory

A monograph by Messarovich and Mako is considered the first mathematical formalization of hierarchical systems [3]. They define hierarchical systems as follows:

1. Single-level single-objective systems are systems in which the goal is defined for the entire system. The main feature of such systems is the absence of contradictions when optimizing the solution.
2. Single-level multi-objective systems are systems in which decision-making lies with agents who are at the same level and have distinct goal-setting, different from the goals of the system. In this case, the occurrence of conflicts between decision makers is not excluded.
3. Multi-level multi-objective systems are systems that represent a hierarchy of subordinates, where agents can have different goal-setting both between levels and each other, and with the entire system.

Multi-level multi-objective systems are of the greatest interest to researchers, since resolving conflicts between the goal-setting of agents and achieving the maximum objective function of the entire system is a complex task. In such systems, the elements of the upper level determine the purposefulness of the lower-level agents, but do not control it [3]. Thus, agents can influence the motivation of lower-level management agents, but they will retain the degree of latitude in decision-making.

In addition, there are several features in hierarchical systems:

1. Upper-level agents see broader aspects of the process associated with a high level of abstraction. The detailing of the process occurs with the transition to the level of lower-level agents gradually. Thus, upper-level agents have a high level of uncertainty.
2. The control actions of the upper-level agents occur with a longer time period as compared with the lower-level ones.
3. The description of the system at the upper level has a high level of uncertainty, so the formalization of decision-making becomes more difficult, especially without going to the description of the lower-level agents.

To solve management problems in multi-level target systems, researchers use game theory to formalize strategies and optimization problems. For instance, it was applied to solve management problems at a two-level production enterprise [6] and typical management tasks [7–10]. These studies consider a specific enterprise with poor standardization.

2.2 Multi-agent Systems

Modern trends in the organizational systems management include the integration of game-theoretic formalization of the system with multi-agent systems. In this case, agent strategies are set for groups of agents, based on which multiple game simulation (imitation) and optimization are performed.

Multi-agent systems are distributed information systems the use of which is hinge on the interaction of agents [11, 12]. The term "agent" in the agent-oriented approach is understood as active, autonomous, communicative, motivated objects [11]. Agents have the ability to communicate and influence both other agents and the entire system [13, 14]. One agent simulates the behavior of one person, and multi-agent systems simulate the behavior of a group of people and even the society [15].

The use of multi-agent systems allows modeling the agents' environment, parameters and patterns of their behavior, dependencies and consequences of their actions. The agent's parameters can be set with the objective f_i $(\cdot)$ and utility functions v_i $(\cdot)$, and the system can be defined with parameters that reflect the "technologies" function w$(\cdot)$. Based on game theory, the properties of the game are set, including the sequence of actions, strategies, etc. Using methods of scenario reduction and optimization, an optimum is achieved, which is a solution in which the objective function of the entire system will be maximal.

The use of multi-agent systems is premised upon the increasing complexity of organizational systems, which does not allow using the approaches applied earlier. This conclusion especially concerns the intuitive decision-making in management. Following the concept of decision-making based on the management object models in most cases allows solving this problem. The aim is to create a model which would be a digital analogue of the management object allowing to conduct experiments and test various scenarios of impact on the system. The developed models have varying precision and reflect the selected aspects of the organizational system functioning, which depends on the application purpose.

2.3 Formalization of Organizational System Management

Organizational system management $(U_A \times U_V \times U_I \times U_R \subseteq U)$ is the impact on it in order to ensure the required behavior, including the result of this behavior.

The impact can be directed to one of the system parameters [16]:

1. Composition; includes agents and system components;
2. Structure; a set of information, management, technological, and other relationships between agents;
3. Set of possible actions;
4. Objective functions of participants reflecting their interest;
5. Awareness; a set of information owned by agents;
6. Functioning order; a sequence of making decisions, actions, obtaining information, etc.

When two or more agents are involved in decision-making, management consists in influencing decision-making by another (related) agent. The agent's decision-making model is a tuple $\Psi = \{A, A_0, \theta, v(\cdot), w(\cdot), I\}$, where A is a set of possible actions, A_0 is a set of permissible results, Θ is a set of possible values of situations (uncertainty), $v(\cdot)$ is a utility function, $w(\cdot)$ is the "technologies" function, I is information. Uncertainty is defined as the agent's awareness of the external and internal system parameters. In this case, uncertainty covers both external and internal one. The external environment $\Theta o \subset \Theta$ includes everything that does not depend on a particular system: environmental conditions, regulations, relation of other systems to the system, etc. The internal uncertainty $\Theta i \subset \Theta$ can include the enterprise's resources, i.a. labour ones, the equipment condition, etc. As part of the uncertainty consideration, it is noteworthy that the information in the decision-making tuple can determine the knowledge of uncertainty parameters. The "technologies" function $w(\cdot)$ reflects the structure of the managed object [16].

Denote the relationship of the performance from the agent's action. When deciding on the impact on the management object, the $y \in A_0$ result will depend on its parameters and the existing uncertainty (internal and external parameters of the system): The $z = w(y, \theta)$ "technologies" function appears, since the agent's actions are directed at the management object structure; the results directly depend on it.

The objective function of the Ai agent is defined through the utility function of the performance: $f_i = v_i(w(y, \theta_i))$. Each management agent has its own utility function, which is determined by motivation to participate in the process and get the result. Agents strive to maximize their objective function. Thus, they will strive to make such a decision: $y_i = Agrmax f_i(y)$.

The goal-setting of the entire system can also be represented through the utility function: $f_0 = v_0(w(y, \theta)$. Consideration of the objective function of the management agents' functions is due to the possibility of applying various hypotheses and types of games when formalizing the organizational system management.

Based on the composition of the decision-making model tuple, the following types of management are distinguished [16, 17]: institutional, motivational, and information. Institutional management ($u_A \in U_A \subseteq U$) is an impact on the activity regulations; thus, the impact occurs on a variety of permissible actions (A). Motivational management ($u_V \in U_V \subseteq U$) is aimed at changing the utility function ($v(\cdot)$). In information management ($u_I \in U_I \subseteq U$), the impact occurs on the information (I) that the agent uses to make decisions. Consider an organizational system management model in which decisions between three levels are distributed according to these decision types.

In addition, the management agents of the organizational system have the ability to influence the flow of resources into the process, which affects the control action on the management object. Thus, we will also introduce the concept of indirect control action on the resources of the process $u_R \in U_R \subseteq U$. This control action affects the internal uncertainty Θ_i.

3   Results

3.1   Decision-Making in a Multi-level Objective System

The three-level system is recognized as simple-basic for multi-level objective systems [4]. Based on the mathematical formalization of such systems, it will be easy to changeover to systems with a large number of management agents. Consider one

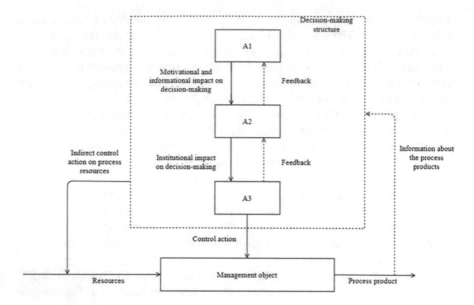

Fig. 1 Management model in a three-level multi-objective system

of the possible options for managing a hierarchical multi-level objective system (Fig. 1). There are three agents in the system (A1, A2, A3), the control between which is distributed, but only A3 produces a direct control action on the process. A typical three-level system has a planning distributed into three levels: strategic, tactical, and operational. Thus, the strategy contains system-wide goals; tactics are methods for achieving these goals, and operational planning consists in choosing specific achieving tools.

In hierarchical multi-level objective systems, decision-making, as well as management, is distributed. The upper levels (A1) of decision-making have a high level of the uncertainty representation abstraction ($\theta_1 \in \Theta$) associated with the process [3]. Therefore, the uncertainty at the upper level of decision-making will be higher or equal (it is possible in some cases of small organizational systems) to the uncertainty at the lower levels (A1, A2, ..., An): $\theta_i \geq \theta_{i+1}$; $\theta_i, \theta_{i+1} \in \Theta, i \in [1, n]$. The transfer of control functions to the lower levels at this point is appropriate due to the reduced level of uncertainty in the decision-making model of the lower-level agent: $\Psi_2 = \{A_2, A_{0_2}, \theta_2, v_2(\cdot), w_2(\cdot), I_2\}$. . In this case, it is required to make a motivational impact on A2 ($u_V \in U_V$), consisting in the message of the target outcomes $Z \subseteq A_0$, where Z is the set of acceptable results of the process from the set A_0, as well as informational impact ($u_I \in U_I$), consisting in the transmission of information about the external environment, parameters of other agents, activity predictions, etc. [16]. It is noteworthy that each agent has a number of preferences on a set of results $Z \subseteq A_0$, while the set Z consists of a set $z = w(y, \theta)$, where $y \in A$ is the impact on the management object.

An institutional impact ($u_A \in U_A$) is made on the subordinate agent (in the model under consideration, it is A3), in which a set $y_3 \in A$ is reported to him or her. At the same time, the agent is no longer informed of a set Z, but of actions in which the upper-level agent A2 expects to receive the targets $z = w(y_3, \theta_3) \in Z \subseteq A_0$ that agent A1 informed him or her about by making an impact u_I. In this case, the result of the activity depends on the actions $y_3 \in A$ of agent A3 and the environment $\theta_3 \in \Theta$.

The objective function of the organizational system in the case determined is specified as follows (1):

$$f_0(y_3) = v_0(w(y_3, \theta))\tag{1}$$

The hypothesis of the agent's rational behavior is that out of all possible options, with the information available, the agent will make a decision that maximizes the total utility [17]. When making a decision, the agent selects an action from $P(y) \subseteq A$. Therefore, for agents A1 and A2, the actions will be to select $u \in U$. As part of the formalization of managerial decision-making in a multi-objective hierarchical system, it stands to mention that the lower-level agents have less idea of the utility function $v_0(w(y))$ of the entire system than the upper-level ones. Moreover, the remoteness of agents from the center leads to high uncertainty $\theta \in \Theta$ in understanding the objective function of the center, which is associated with the benevolence hypothesis. Thus, the hypotheses of rational behavior and benevolence in a multi-level system are inapplicable.

In addition, agents with a greater understanding of the objective function of the system (located higher in the hierarchy) have an indirect influence on the impact on the management object y. Thus, when deciding on the control action on the lower-level agents, it is necessary to be aware of the objective function $f_i(y)$ of agent Ai.

3.2 Oil Production Enterprise

Consider the formalization of a hierarchical multi-objective organizational system on the example of an oil production enterprise. The oil production enterprise is a typical representative of the organizational system with the management distributed into three levels: the parent organization, subsidiary, and administrative and managerial personnel (AMP) of the field (Figs. 2 and 3).

The parent organization is responsible for following the strategies of the company and the state, which also influences the process, but not the managerial one. Thus, its goal-setting level is strategic. A subsidiary is subordinate to the parent organization, while having a certain degree of latitude. At this level, decisions are made based on the requirements and restrictions $li \in Li$ provided by the upper-level agent. The budget is distributed between subsidiaries depending on a variety of indicators: the feasibility of developing fields managed by them; the area of fields; possible activities

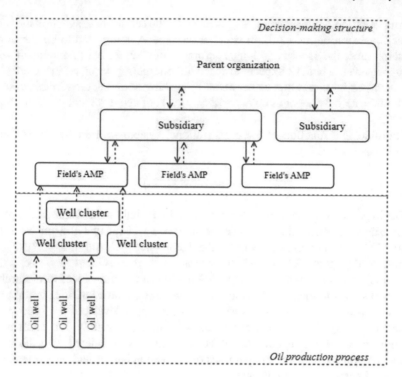

Fig. 2 Hierarchical management of the oil production enterprise

to be carried out; the number of personnel, etc. Therefore, the control action of the parent organization to the subsidiary is motivational $u_V \in U_V$, presented in the form of submitted requirements li for technological and economic indicators of production, and informational $u_I \epsilon U_I$, presented in the form of restrictions on the input flows of the oil production process and information about the external and internal environment $\theta_o, \theta_i \in \Theta$.

The subsidiary has an institutional impact on the field's AMP, which consists in transmitting the plan and project for the field development. In these conditions, the AMP can adjust plans and perform operational management, which has not been originally incorporated in the plan. The input flows of the oil production process $u_R \in U_R$ are managed by both the parent organization and subsidiary through step-by-step flow diversification. In this way, the parent organization determines the input flows of information, material, and financial resources for subsidiaries that combine fields. The second agent decides on the allocation of resources between the fields in accordance with the plan that was selected based on motivational u_V and informational u_I impacts.

The feedback information flows (I) contain, among other things, the decisions made and the results $z \in A_0$, and are directed to the field's AMP for analysing, accounting, operational decision-making, etc. The sum of these flows $\sum_{i=1}^{M} I$ on

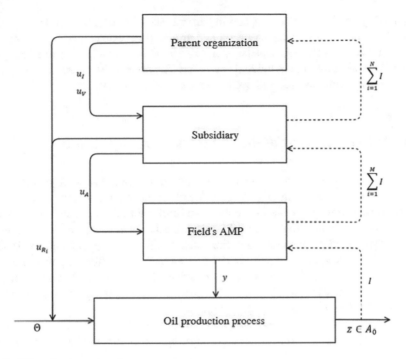

Fig. 3 Oil production enterprise management

fields go to subsidiaries, where they are aggregated and directed to the parent organization $\sum_{i=1}^{N} I$.

When making a decision on performing a control action, the parent organization focuses on the requirements imposed by market conditions, the strategies of the state and the company. Uncertainty $\theta_1 \in \Theta$ is limited by external environment and aggregated input streams. The information I based on which subsidiaries make decisions is limited by detailed input flows of field resources, but the organization has incomplete information about the external situation and the actions of other agents.

Although A1 has a motivational impact on A2, it determines only the output flows (results) of the process $z \in A_0$. In addition, the subsidiary has its own motivations for the distribution of teams, the process of field development, including the closure and commissioning of new wells. Given a certain number of subsidiaries, the parent organization cannot have full information about the objective function of A2. At the same time, A2 has full information about A3's actions in terms of tactical decisions; however, some of them may come with a time lag.

One of the main problems of managing multi-objective hierarchical systems is the complexity of the management process and the associated difficulties in analyzing its functioning [2]. In this regard, the management process is also complicated, including the control action on agents and actions aimed at the process. In addition, the classical cybernetic model implies the impact formed by the management subject on the managed system is not specified, and in general takes the form of intuitive decision-making, or decision-making based on accumulated experience, or other available management patterns. This puts the management object's functioning efficiency in a strong dependence on the individual characteristics of the subject, and actually takes this object far from optimal and even quasi-optimal scenarios.

3.3   SM-Based Decision-Making. Multi-agent Systems

Delegation of control functions is an integral part of the hierarchical objective system [4]. In this case, the agents of the upper levels of management become dependent on the actions of the lower-level agents when implementing both their own function and the aggregate objective one. Multi-agent simulation systems are widely applied to make informed management decisions in a multi-objective system [15, 18–20], which allow designing an environment and agents with different goals. To solve the problems of optimizing managerial functions in a multi-objective hierarchical system, using the example of an oil production enterprise, we propose to develop a conceptual model of a multi-agent system.

Based on the cybernetic model (Fig. 3), we have developed a model for the implementation of managerial functions at an oil production enterprise using a multi-agent simulation model (Fig. 4).

Simulation models allow building simulations of the system's state depending on the set parameters, which is predictive analytics [21]. The parent organization, subsidiary, and field's AMP were identified as agents of the multi-agent system. Management objects are represented as a set of lower-level agents, which is associated with the representation of an entire hierarchical system. In this manner, the agent of the parent organization simulates the impact on the management object at the level of subsidiaries, which is represented by all organizations subordinate to the agent, as well as fields' AMP subordinate to upper-level agents, and the process of oil production itself. The parent organization makes hypothetical impacts on the management object $u_{Ih}, u_{Vh}, u_{R1h} \in U$, while receiving information $\sum_{i=1}^{N} I_h$ about the outputs of the oil production process $\sum_{i=1}^{N} z_h$. Focusing on the organizational system's indicators, agents can adjust their impact $u \in U$. Similarly to the above description, the subsidiary agent and the field's AMP act.

The use of a multi-agent system allows simulating behavior (control action $u \in U$) of agents with different goal-setting in accordance with the specified parameters (management tuples Ψ), as well as their impact on the external environment $\theta_o \in \Theta$, which is a component of other agents' tuples.

A simulation model application gives the management subject the opportunity to evaluate the results of the system's response to the control action. In this case, it will have an analytical justification for the decisions taken. Simulation models with stochastic events, an integral part of social systems, predict pessimistic and optimistic scenarios, which also makes it possible to assess and manage risks.

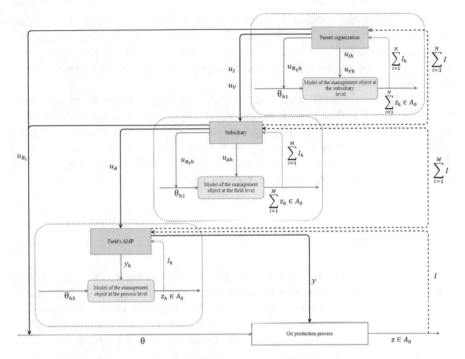

Fig. 4 Conceptual model of the application of the multi-agent system for managing a multi-objective hierarchical organizational system

4 Discussion

Undergoing digital transformation processes, including the Industry 4.0 revolution, enterprises produce a large amount of data that can potentially be used in making management decisions [17]. Researchers are actively investigating the application of big data for decision-making [23]. They develop specialized technologies for data acquisition, processing, [23, 24] and analysis [23, 25–27]. Among the issues of scientific interests are the following: the rationality of decisions and the impact on the entire system effectiveness [28], mathematical formalization of the decision-making process [29, 30], and the use of simulation models [28, 29].

With the growth of digital data amount, a sociotechnical system becomes difficult to formalize and structure. This is due to the emergence of new dependencies that can be taken into account when building a system model. Hybrid models are increasingly being used in the systems design [31]. Hybrid modeling is a modeling which uses several paradigms when building a model. The advantages of its application include the data availability. It is achieved by combining several sources for various paradigms due to different levels of modeling abstraction [32]. Data can be represented as digital data collected from sensors, etc. [22].

The use of digital data has both advantages and disadvantages, which are described in detail in [33]. The main advantages of utilizing Industry 4.0 technologies for enterprises are the data relevance, accuracy, and timeliness. This is particularly relevant when compiling dynamic models to support managerial decision-making in organizational systems.

However, digital transformation also carries some difficulties associated with the new context of this data [33]. It is also worth paying attention to the features of digital data acquisition and processing, which are becoming more complex and more expensive in comparison with non-digital ones. If it is the use of digital data types that is evaluated, then it does not entail major drawbacks. This is becoming the driver for the intense digital transformation of organizations.

5 Conclusion

The chapter considers the problem of making informed management decisions in organizational systems, as well as the possibilities and requirements for information systems to support decision-making. The result of the work is a conceptual model of a three-level multi-objective system. The study describes the game-theoretic formalization of management, including control actions performed by agents, feedback flows, agents' objective functions, i.a. utility functions. Based on the provided formalization, the chapter describes the organizational system of an oil production enterprise, which is a typical three-level enterprise divided into decision-making levels: strategic, tactical, and operational. By analyzing the system, we developed a conceptual management model at an oil production enterprise using a multi-agent system. Within the developed system, agents are represented as decision-making organizations: parent organization, subsidiary, administrative and managerial personnel of the field.

The research will continue in two directions. The first one is to detail the structure of the system and the decision-making tuple, as well as strategies and sequences of agents' actions. The second one is to develop algorithms for the multi-agent system intelligent module, which will optimize the games in the system.

Acknowledgements The research is funded by the Ministry of Science and Higher Education of the Russian Federation (contract No. 075-03-2023-004 dated 13.01.2023).

References

1. Reshitko MA, Ougolnitsky GA, Usov AB (2023) Numerical method for finding nash and shtakelberg equilibria in river water quality control models. Comput Res Model 12(3): 653–667. (June 2023). https://doi.org/10.20537/2076-7633-2020-12-3-653-667
2. Mishina NS (2018) Problems of decision-making in hierarchical systems. In: Proceedings of the XLVII scientific and educational conference of ITMO university, Federal State Autonomous Educational Institution of Higher Education "ITMO National Research University", St. Petersburg, pp 90–93
3. Mesarović MD, Macko D, Takahara Y (1970) Theory of hierarchical, multilevel, systems. mathematics in science and engineering : a series of monographs and textbooks. Academic Press
4. Novikov DA (1999) Mechanisms of functioning of multilevel organizational systems. Fond "Problemy upravleniya", Moscow, Russia, pp 161
5. Tarasov VB (1998) Agents multi-agent systems, virtual communities: strategic direction in computer science and artificial intelligence. In: Artificial intelligence news, pp 5–63
6. Wernz C, Deshmukh A (2007) Decision strategies and design of agent interactions in hierarchical manufacturing systems. J Manuf Syst 26(2):135–143. https://doi.org/10.1016/j.jmsy.2007.10.003
7. Wernz C, Deshmukh A (2007) Managing hierarchies in a flat world. In: Proceedings of the 2007 industrial engineering research conference, Nashville, TN, pp 1266–1271
8. Wernz C, Deshmukh A (2009) An incentive-based, multi-period decision model for hierarchical systems. In: Proceedings of the 3rd annual conference of the Indian Subcontinent Decision Sciences Institute Region (ISDSI), Hyderabad, India, pp 12–17
9. Wernz C, Deshmukh A (2010) Multi-time-scale decision making for strategic agent interactions. In: Proceedings of the 2010 industrial engineering research conference, Cancun, Mexico, pp 1–6
10. Wernz C, Deshmukh A (2010) Multiscale decision-making: bridging organizational scales in systems with distributed decision-makers. Eur J Oper Res 202(3):828–840. https://doi.org/10.1016/j.ejor.2009.06.022
11. Barambones J, Imbert R, Moral C (2021) Applicability of multi-agent systems and constrained reasoning for sensor-based distributed scenarios: a systematic mapping study on dynamic DCOPs. Sensors 21(11):3807. https://doi.org/10.3390/s21113807
12. Vistbakka I, Troubitsyna E (2021) Modelling resilient collaborative multi-agent systems. Computing 103:535–557. https://doi.org/10.1007/s00607-020-00861-2
13. Nair AS, Hossen T, Campion M et al (2018) Multi-agent systems for resource allocation and scheduling in a smart grid. Technol Econ Smart Grids Sustain Energy 3:15. https://doi.org/10.1007/s40866-018-0052-y
14. Samigulina G, Samigulina Z (2020) Ontological model of multi-agent Smart-system for predicting drug properties based on modified algorithms of artificial immune systems. Theor Biol Med Model 17:12. https://doi.org/10.1186/s12976-020-00130-x
15. Canese L, Cardarilli GC, Di Nunzio L et al (2021) Multi-agent reinforcement learning: a review of challenges and applications. Appl Sci 1:11
16. Novikov DA (2004) Institutional management of organizational systems. IPU RAN, Moscow, Russia, pp 68
17. Gubko MV, Novikov DA (2005) Game theory in the management of organizational systems. IPU RAN, Moscow, Russia
18. Ponnambalam SG, Janardhanan MN, Rishwaraj G (2021) Trust-based decision-making framework for multiagent system. Soft Comput 25(11):7559–7575. https://doi.org/10.1007/s00500-021-05715-3
19. Pan J (2022) Structural optimization of architectural environmental art design based on multiagent simulation system. Math Probl Eng 1–9. https://doi.org/10.1155/2022/4341816
20. Riekki J, Mämmelä A (2021) Research and education towards smart and sustainable world. IEEE Access 9:53156–53177. https://doi.org/10.1109/ACCESS.2021.3069902

21. Bolsunovskaya MV, Gintciak AM, Burlutskaya ZV, Petryaeva AA, Zubkova DA, Uspenskiy MB, Seledtsova IA (2022) The opportunities of using a hybrid approach for modeling socio-economic and sociotechnical systems. In: Proceedings of Voronezh State University. Series: systems analysis and information technologies, vol 3, pp 73–86. https://doi.org/10.17308/sait/1995-5499/2022/3/73-86
22. Anumbe N, Saidy C, Harik R (2022) A primer on the factories of the future. Sensors 22(15):5834. https://doi.org/10.3390/s22155834
23. Duan L, Da Xu L (2021) Data analytics in Industry 4.0: a survey. Inf Syst Front. https://doi.org/10.1007/s10796-021-10190-0
24. López-Ballesteros A, Trolle D, Srinivasan R, Senent-Aparicio J (2023) Assessing the effectiveness of potential best management practices for science-informed decision support at the watershed scale: The case of the Mar Menor coastal Lagoon, Spain. Sci Total Environ 859:160144. https://doi.org/10.1016/j.scitotenv.2022.160144
25. Bousdekis A, Mentzas G (2021) Enterprise integration and interoperability for big data-driven processes in the frame of Industry 4.0. Front Big Data 4:644651. https://doi.org/10.3389/fdata.2021.644651
26. Meenakshi N, Kumaresan A, Nishanth R, Kishore Kumar R, Jone A (2023) Stock market predictor using prescriptive analytics. Mater Today Proc 80:2159–2166. https://doi.org/10.1016/j.matpr.2021.06.153
27. Menezes BC, Kelly JD, Leal AG, Le Roux GC (2019) Predictive, prescriptive and detective analytics for smart manufacturing in the information age. IFAC-PapersOnLine 52(1):568–573. https://doi.org/10.1016/j.ifacol.2019.06.123
28. Polhill JG, Edmonds B (2023) Cognition and hypocognition: discursive and simulation-supported decision-making within complex systems. Futures 148:103121. https://doi.org/10.1016/j.futures.2023.103121
29. Kikuchi T, Kunigami M, Terano T (2023) Agent modeling, gaming simulation, and their formal description. In: Kaihara T, Kita H, Takahashi S, Funabashi M (eds) Innovative systems approach for facilitating smarter world. Design science and innovation. Springer, Singapore. https://doi.org/10.1007/978-981-19-7776-3_9
30. Grosz BJ, Kraus S, Talman S, Stossel B, Havlin M (2004) The influence of social dependencies on decision-making: initial investigations with a new game. In: Proceedings of the third international joint conference on autonomous agents and multiagent systems, 2004. AAMAS 2004, New York, NY, USA, 2004, pp 782–789
31. Burger K, White L, Yearworth M (2019) Developing a smart operational research with hybrid practice theories. Eur J Oper Res 277(3):1137–1150. https://doi.org/10.1016/j.ejor.2019.03.027
32. Lattila L, Hilletofth P, Lin B (2010) Hybrid simulation models-When, Why, How? Expert Syst Appl 37:7969–7975
33. Gintciak A, Burlutskaya Z, Fedyaevskaya D, Budkin A (2023) Use and processing of digital data in the era of Industry 4.0. In: Ilin I, Petrova MM, Kudryavtseva T (eds) Digital transformation on manufacturing, infrastructure & service. DTMIS 2022. Lecture notes in networks and systems, vol 684. Springer, Cham. https://doi.org/10.1007/978-3-031-32719-3_36

Digital Modelling of the System of Knowledge Exchange and Building Within a Network of Industrial Enterprises

Zhanna V. Burlutskaya⬦, Aleksei M. Gintciak⬦, and Lo Thi Hong Van⬦

Abstract The chapter aims to the study of the opportunities of using simulation modeling to describe the system of the knowledge exchange and building within a network of industrial enterprises. Such systems are responsible for the innovative development of companies, ensuring the development of more functional and resource-intensive products. Thus, innovative activity leads to sustainable production and consumption by ensuring high consumer value of goods. Within the framework of this work, the features of building industrial enterprise networks are considered from the point of view of possible strategies for combining companies. Based on the information received, the selection of simulation modeling tools is given. The result of the work is a conceptual hybrid model of the system of knowledge exchange and building within a network of industrial enterprises. The hybrid model being developed combines such paradigms of simulation modeling as system dynamics and agent modeling. Agent modeling tools are used to describe the dynamics of networks of industrial companies formed from independent agents–companies. System dynamics is used to describe the dynamics of knowledge flows and their economic equivalent. The results of the simulation of the agent model determine the coefficients that affect the processes of knowledge building. In turn, the choice of strategies by agents depends on the dynamics of accumulated knowledge and their financial equivalent. This work is an analytical basis for the development of a hybrid simulation model of the system of knowledge exchange and building within a network of industrial enterprises.

Keywords Digital transformation · Knowledge management · Operations research · Simulation modelling · Decision support systems

Z. V. Burlutskaya (✉) · A. M. Gintciak
Peter the Great St. Petersburg Polytechnic University, St. Petersburg, Russian Federation
e-mail: zhanna.burlutskaya@spbpu.com

L. T. H. Van
University of Economics and Business, Vietnam National University, Hanoi, Vietnam

A. Bencsik and A. Kulachinskaya (eds.), *Digital Transformation: What is the Company of Today?*, Lecture Notes in Networks and Systems 805,
https://doi.org/10.1007/978-3-031-46594-9_3

1   Introduction

A lot of scientific and analytical works deal with innovation management, each of which shows certain aspects of the knowledge exchange and building process peculiarities. Part of the studies focuses on the knowledge diffusion both between enterprises and within the region [1, 2]. Others consider approaches to building knowledge through collaborative research [3–8]. In this case, relations of the following types are examined: research enterprises–industrial enterprises, industrial enterprises–industrial enterprises, etc. However, it was not possible to find studies that consider the processes of knowledge exchange and building within a single integrated model or a set of interrelated models. However, the issues of the dissemination of knowledge and the mechanisms of their increase are of high importance both for the scientific community and for society. Technological innovations lead to the development of more functional and resource-intensive products with higher quality. Thus, the consumer value of an individual product increases by improving quality and reducing costs. The latter is true provided that the more innovative product is an improved equivalent of two or more existing products. In such a paradigm, technological innovations become a tool for the transition to sustainable consumption and production.

However, the realization of the innovative potential of an individual company, an industrial cluster or a region as a whole requires additional resources, capabilities and skills that can be acquired by integrating research activities with other companies. On the other hand, large industrial companies, which are a complex system consisting of many branches, often lose knowledge at the stage of spreading new technologies within the company. So, the larger the enterprise or network of enterprises, the more complex the processes of knowledge transfer and joint development.

This work is aimed at solving the global problem of organizational management of knowledge flows between interconnected agents within a single system.

Knowledge diffusion models are considered within strictly defined static networks. Such a division is conditioned by the complexity of describing socioeconomic systems with a distinctive low determinism and complexity of processes. Hybrid simulation can be used to solve these problems. Simulation modeling allows exploring the system in a secure environment of digital technologies [9]. A kind of a real system digital analogue is created with a certain degree of abstraction. The obtained result is calibrated with regard to the available real data on the behavior of the system, and then, using mathematical prediction tools, the behavior of the system is simulated depending on changes in internal or external parameters. Such a system makes it possible to describe social systems with high uncertainty. However, a single simulation model will not allow us to consider all aspects of knowledge exchange and building. In this case, we apply a hybrid approach [10]. It implies two or more simulation models implemented within different paradigms of simulation modeling being built. Each of the paradigms presents its own perspective on the system.

In this connection, the present chapter aims to develop the prototype of a conceptual hybrid model of knowledge exchange and building within a network of industrial enterprises. Within the research, the features of building industrial enterprise networks are considered from possible strategies for merging enterprises, as well as their strengths and weaknesses perspectives. Based on the information received, the selection of simulation modeling tools is given, taking into account their application specifics. The research result is a conceptual hybrid model of the system of knowledge exchange and building in the industrial enterprise network.

2 Materials and Methods

2.1 Review of the Specifics of Building Industrial Enterprise Networks

Most of the analyzed articles (about 70%) deal with single-cluster enterprise networks. The related literature review has shown that the industrial cluster success de-pends on a wide network of connections at both local and cross-border levels, as well as its management for access to relevant knowledge and resources [6]. Thus, in a wide cluster network, unique local knowledge of each enterprise is retained separately and then they are supplemented by other enterprises' similar experience. In such cluster networks, a node or hub enterprise is often allocated. In this case, this enterprise is considered as the network innovation capacity center and is responsible for the cluster evolution. Since not all firms have the same qualifications for the introduction and use of new knowledge, it is often the concentrator firm that has the greatest tendency to identify and absorb new knowledge. It is noteworthy that regardless of the individual innovation potential of the hub enterprise or other network enterprises, the cluster will be weakened without the synergy of resources and well-established knowledge acquisition processes [5].

However, there are proponents of the opinion that international expansion undermines the cluster technological advantage and core industrial values such as local knowledge and research infrastructure [7]. Moreover, while clusters help provide knowledge advantages, they can also potentially hinder investment [4]. Since clusters also include a large group of competitors and related enterprises, they are places of intensive, fierce competition, and high production costs. In addition, if knowledge were really localized in separate industrial clusters, the landscape of the global knowledge economy would be a world of local knowledge pools [5].

Another approach to networking is based on partnerships between enterprises participating in a single value chain. Thus, enterprise networks may include enterprises from similar but different types of activities. In this way, innovation will be stimulated by a combination of related but differentiated knowledge. It is crucial to mention that unrelated knowledge is of less value to enterprises [4]. This approach solves the problem of high competition within the network, but requires more attention to the knowledge transfer structure from both a conceptual and empirical perspectives [4].

Another attribute of the network success is the availability of financial resources. When considering the success story of Silicon Valley, one may notice the connection between the technological development of the region and the development of the California venture industry [8].

Based on all of the above, several conclusions can be drawn. Firstly, enterprise network models can be divided into: models of networks of companies from one cluster, models of networks of companies from interconnected clusters, models of networks of companies from complementary clusters and models of networks of companies with international connections. These models may be hybrid.

Secondly, the success of innovative development directly depends on the strength of connections within the network. Thus, the passivity of individual companies can lead to the "burning" of the innovative potential of the entire network.

Finally, the importance of financial support for innovative developments should not be underestimated. In this case, financial support can be obtained both from the resources of the network and through venture capital.

2.2   Application of Simulation Modeling for the Socioeconomic Systems Analysis

Please note that the first paragraph of a section or subsection is not indented. The first paragraphs that follows a table, figure, equation etc. does not have an indent, either.

Subsequent paragraphs, however, are indented. Comparing technical, sociotechnical, and socioeconomic systems, one can notice a decrease in the level of determinism and a concomitant increase in complexity with an increase in the proportion of social agents [9]. As a result, the socioeconomic system will have a considerable uncertainty, expressed in the unpredictable dynamics of the system and the its agents' behavior. However, there are numerous studies aimed at solving the problem of making informed management decisions in the management of socioeconomic systems. Simulation modeling is the most flexible tool for studying socioeconomic systems, combining mathematical logic with empirical re-search [11, 12].

Simulation modeling is applied when studying complex systems based on their digital counterpart exploration. Thus, the simulation model repeats the object under study with a certain degree of detail and accuracy [9]. Such a model becomes a source of additional information about the behavior of the system by conducting simulation experiments. The results of the experiments allow us to answer the question of system functioning peculiarities depending on changes in internal or external conditions. Since experiments on a real socioeconomic system are nearly impossible, simulation modeling is in fact the only tool for predicting the system dynamics. Unlike expert assessment methods, simulation modeling has greater accuracy due to the use of formalized mathematical prediction methods. Moreover, predictions based on expert opinion are used only in the short term, while mathematical predictions are used on all planning horizons.

An additional advantage of applying simulation modeling is the dynamically developing base of diverse approaches and tools. For instance, various simulation modeling methods are used for systems with different levels of detail and features of the simulated process [11]. Thus, the behavior of the demographic system can be considered with both system dynamics and agent-based modeling. In the first case, the system will be described as a set of interconnected flows with a high degree of abstraction [9]. The results obtained will allow us to assess global trends affecting demographic flows. At the same time, considering the system as interconnected agents, we get a high degree of detail and analysis of the strategies of each agent or agent groups. Such models allow assessing the impact of particular events on the selected social group. In addition to system dynamics and agent modeling, discrete-event simulation, system dynamics, and game-like simulation are also distinguished. How-ever, we will not consider them in this work, since they are more appropriate for high-ly specialized industries or sociotechnical systems.

Despite the advantages of each simulation type, hybridization is the most promising approach [12–16]. The hybrid approach allows considering the system from several perspectives at once and at different levels of abstraction [13, 15, 16]. In this way, it can be either a sequential application of models for study with varying degrees of detail, or parallel modeling, which allows exchanging results at selected steps and therefore refine the model step by step.

Consider the application of a hybrid approach with regard to the problem being solved. There are a number of enterprises with a certain amount of knowledge [15]. On the one hand, it is the agent interaction of enterprises with each other. As part of their interaction, enterprise agents decide whether to enter or exit the corresponding enterprise network. What is more, enterprises make decisions on the knowledge transfer, financial support for scientific and technical developments, or participation in training. Thus, there is a clear strategic interaction between network agents. On the other hand, knowledge is represented with flows that have the ability to accumulate and change in some conventional units. In turn, the accumulated knowledge can be translated into a financial equivalent that allows companies to continue their innovative activities.

2.3   Hybrid Modelling

In the last decade, hybrid modeling has been developed in operations research. This term refers to a combination of different paradigms in order to model a single system.

Hybrid modeling is a modeling that combines two or more modeling methods to achieve results that cannot be obtained using combined methods separately.

It should be noted that in this case we are not talking about using different approaches to modeling within one model or even within one program tool. Hybrid modeling is an exclusively methodological approach that allows studying the same simulated object from different sides or at different levels of abstraction. However, the creation of software designed specifically for hybrid simulation is possible in the very near future.

Although the concept of combining different types of modeling is not particularly complex and original, in recent years its popularity has continued to grow. There are reasons to believe that this fact is associated with the need for a more complete study of the systems of interest using separate modeling approaches.

The popularity of the hybrid approach is also increasing when modeling sociotechnical systems for management optimization–this is confirmed by the positive dynamics of the number of publications in the domains of operations research, control systems, and industrial engineering.

The transition to hybrid modeling allows to get much more information about the system, considering it from the point of view of different elements, which often do not overlap within the framework of classical simulation approaches. However, the question of choosing a certain approach and type of modeling causes additional difficulties due to the excess of possible options.

As part of the solution of the designated task, the hybrid approach will look at the system of knowledge dissemination from two points of view. So agent-based modeling will be used to describe the behavior of companies in the context of innovation management. Such a model will allow us to assess the prospects for the development of companies depending on the activity and desire for unification. The main process of knowledge dissemination will be described by system dynamics. Thus, the models will be dependent on each other.

3 Results

The result is that there are two independent models: an agent-based model of enterprise interaction and a system-dynamic model of knowledge dynamics. Each of them considers the system from a certain perspective and can operate separately. However, the initiating model is the agent model. Since the knowledge flows correspond to each enterprise network, the accumulated knowledge value will be zero at the initial time. With an agent, the accumulated knowledge value will be equal to this agent's accumulated knowledge value.

Let's consider a hybrid model of the knowledge exchange and building system in a network of industrial companies (Fig. 1).

The appearance of the agent initiates the simulation. Each agent can be in one of the states visualized on Statecharts. At the stage of the agent's transition to the "agent of the network" state, the model switches to the active state. This means that any further action of the agent sets in motion a model of system dynamics. So, when switching to the "active agent" state, the process of accumulation of knowledge begins, the increase of which is a function of the knowledge value of this agent. In turn, the accumulated knowledge is transformed into financial resources. The next iteration

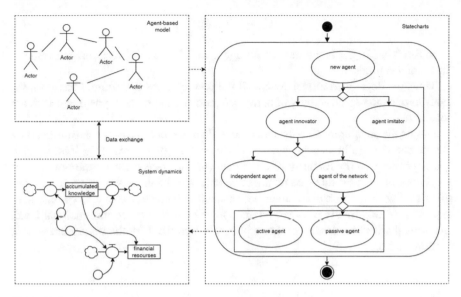

Fig. 1 Hybrid model of the system of knowledge exchange and building within a network of industrial enterprises

of the agent model is performed taking into account the results of the simulation of the system dynamics model and updated data on accumulated knowledge and financial resources.

In this case, we will have a certain set of agents A, in which each agent corresponds to a certain set of states (we presume that these are: new agent, agent innovator/agent imitator, independent agent/agent of the network, active agent/passive agent), some set of parameters and some set of functions.

$$A = \{C; P; F\} \tag{1}$$

In formula (1) C is a set of states, then C_i is one of the possible states of the agent. Accordingly, P is the set of agent parameters, and F is the set of agent functions.

It is assumed that a change in the agent's state affects the accumulated knowledge value $K(t)$.

$$K(t) = K(t - dt) + (K) \times dt \tag{2}$$

In the formula (2), $K(t)$ is the accumulated knowledge for a certain period of time.

In turn, the dynamics of accumulated knowledge $K(t)$ affects the financial resources of companies M. At this stage, feedback arises from the system dynamics model to the agent model. This relationship is expressed by the fact that the accumulated value of financial resources directly affects the agent's choice of behavior strategy.

$$C_{ij} = f(M(t)) \tag{3}$$

In formula (3), C_{ij} is the state of the agent, and $M(t)$ is the accumulated value of financial resources.

Thus, the developed model covers all the prerequisites determined based on the analysis of sources: the activity of network agents, financing and type of network are considered.

During the development of the model, it will also be necessary to take into account other factors that influence the agent's choice of strategy. For example, it is assumed that depending on the maturity of the model (maturity here means the presence of a network of at least three companies), agents will form networks with relationships first within one cluster, then between interconnected clusters, complementary clusters, and finally between interconnected clusters at the international level. Accordingly, it is the last level that will be characterized by the maximum benefit for companies.

4 Discussion

Innovation and technology diffusion is associated with a certain competitive advantage loss risk for business. This leads to the isolated development of technologies in private commercial or public centers without the possibility of extension. Existing cross-licensing measures between enterprises do not fully support the innovation diffusion. Even large industrial enterprises, which are a complex multiple-branched integrated systems, often lose knowledge at the stage of diffusing emerging technologies within the enterprise. Therefore, the larger the enterprise or enterprise network, the more complex the knowledge transfer and joint development processes [2].

At the same time, technological innovations lead to the development of more functional and resource-intensive high-quality products. Thus, the consumer value of a certain product increases by improving quality and reducing costs. The latter is true provided that the more innovative product is an improved equivalent of two or more existing products. In this paradigm, technological innovations become a tool for the transition to sustainable consumption and production.

Thus, large enterprises concerned about the knowledge diffusion through the internal networks of the enterprise's branches can utilize the developed model. Moreover, at the strategic planning stage, the developed model will allow identifying enterprise expansion opportunities, taking into account the possible knowledge integration with other industrial enterprises.

Since the developed model of the system of knowledge exchange and building within a network of industrial enterprises has a complex structure, at the next stage of its implementation, an agent-based model of interaction between enterprises will be developed, taking into account strategies for connecting, exiting, or internal collaboration with other enterprises within the network. This decision stems from the fact that it is the agent model that initiates the system dynamics model describing the dynamics of knowledge flows. It is worth noting that in the agent model being developed, it is assumed to divide agents into four groups according to the Gartner quadrant. It is assumed that depending on the position of the company in the Gartner quadrant, preferences in choosing a strategy will change [17]. For example, leaders are more likely to strive to stimulate innovative developments, sparing no financial resources.

5 Conclusion

Innovation and technology diffusion is associated with a certain competitive advantage loss risk for business. This leads to the isolated development of technologies in private commercial or public centers without the possibility of extension. Existing cross-licensing measures between enterprises do not fully support the innovation diffusion. Even large industrial enterprises, which are a complex multiple-branched integrated systems, often lose knowledge at the stage of diffusing emerging technologies within the research will continue in two directions. The first one is to detail the structure of the system and the decision-making tuple, as well as strategies and sequences of agents' actions. The second one is to develop algorithms for the multi-agent system intelligent module, which will optimize the games in the system. The chapter aims to the study of the opportunities of using simulation modeling to describe the system of exchange and knowledge building in the network of industrial companies. In the course of the study, the features of the association of companies in the network for the purpose of joint research and development are considered. Based on the analyzed information, simulation modeling tools relevant to the description of the selected system are selected. The results obtained are used in the development of a conceptual hybrid model of the system of knowledge exchange and building in a network of industrial companies. The developed hybrid model combines the advantages of such paradigms of simulation modeling as system dynamics and agent modeling. Agent modeling tools are used to describe the dynamics of networks of industrial companies formed from independent agent companies. System dynamics is used to describe the dynamics of knowledge flows and their economic equivalent. Thus, the simulation results of the agent model determine the coefficients that affect the processes of knowledge building. In turn, the choice of strategies by agents depends on the dynamics of accumulated knowledge and their financial equivalent.

This work is an analytical basis for the development of a hybrid simulation model of a knowledge exchange and building system in a network of industrial companies.

Acknowledgements The research is funded by the Ministry of Science and Higher Education of the Russian Federation (contract No. 075-03-2023-004 dated 13.01.2023).

References

1. Bellini E, Era CD, Verganti R (2012) A design-driven approach for the innovation management within networked enterprises. In: Methodologies and technologies for networked enterprises, pp 31–57.
2. Ye D, Zheng L, He P (2021) Industry cluster innovation upgrading and knowledge evolution: a simulation analysis based on small-world networks. SAGE Open 11
3. Sassanelli C, Terzi S (2022) Building the value proposition of a digital innovation hub network to support ecosystem sustainability. Sustainability 14:11159
4. Li P, Bathelt H (2021) Location strategy in cluster networks. J Int Bus Stud 49:967–989
5. Ye D, Wu YJ, Goh M (2020) Hub firm transformation and industry cluster upgrading: innovation network perspective. Manag Decis 58(7):1425–1448
6. Turkina E, Van Assche A, Kali R (2016) Structure and evolution of global cluster networks: evidence from the aerospace industry. J Econ Geogr 16(6):1211–1234
7. Turkina E, Van Assche A (2018) Global connectedness and local innovation in industrial clusters. J Int Bus Stud 49:706–728
8. Ferrary M, Granovetter M (2009) The role of venture capital firms in Silicon Valley's complex innovation network. Econ Soc 38(2):326–359
9. Bolsunovskaya MV, Gintciak AM, Burlutskaya ZV, Petryaeva AA, Zubkova DA, Uspenskiy MB, Seledtsova IA (2022) The opportunities of using a hybrid approach for modeling socio-economic and sociotechnical systems. In: Proceedings of Voronezh State University. Series: Systems Analysis and Information Technologies, vol 3, pp 73–86
10. Gu Y, Kunc M (2019) Using hybrid modelling to simulate and analyse strategies. J Model Manag 15(2):459–490
11. De Paula Ferreira W, Armellini F, De Santa-Eulalia LA (2020) Simulation in industry 4.0: a state-of-the-art review. Comput Ind Eng 149:106868
12. Brailsford S, Eldabi T, Kunc M, Mustafee N, Osorio AF (2018) Hybrid simulation modelling in operational research: a state-of-the-art review. Eur J Oper Res
13. Mittal A, Krejci CC (2017) A hybrid simulation modeling framework for regional food hubs. J Simul 1–16
14. Mustafee N, Powell JH (2018) From hybrid simulation to hybrid systems modelling. In: Winter Simulation Conference (WSC), Gothenburg, Sweden, pp 1430–1439
15. Tsvetkova NA (2017) Simulation modeling the spread of innovations Saint Petersburg, Russia. In: International conference on soft computing and measurements (SCM). IEEE, pp 675–677
16. Barbosa C, Azevedo A (2017) Hybrid simulation for complex manufacturing value-chain environments. Proc Manuf 11:1404–1412
17. Chang M-H, Harrington JE (2007) Innovators, imitators, and the evolving architecture of problem-solving networks. Organ Sci 18(4):648–666

Digital Ecosystems Development in Russian Media Industry as a Result of Their Digital Transformation

Arkady Kiselev◉ and Liubov Silakova◉

Abstract In modern world media companies face serious challenges. From one hand, new technological companies enter the media market and displace traditional mass media. From other hand there is an unstable situation in the advertising market in Russia, whereas advertising is the main source of revenue for traditional mass media companies. Authors have an opinion that digital ecosystem creation as a result of digital transformation will help to overcome these difficulties. There are digital ecosystems, which include media services in Russia, however no studies have been found that assess their development. Comparison of media segments was conducted and digital ecosystem development was assessed in this paper. Ecosystems were analyzed by three blocks: ecosystem development as a whole, media segment development in ecosystem and media services development inside media segment. Criteria groups were defined for each block for comparative analysis. Authors make a conclusion, that majority of the biggest traditional media companies in Russia do not have developed digital ecosystem and focus on the development of the main business.

Keywords Digital transformation · Digital ecosystem · Digital maturity · Ecosystems' media segment · Media company · Video and audio streaming · Content

1 Introduction

Nowadays media industry faces different problems, which have negative impact on media companies results. Among them is competitiveness maintaining in the context of rapid development of media products and technologies in response to changes in user preferences. The biggest players use different methods and tools to attract and retain customers, however user preferences change more and more each year. This problem is exacerbated by the emergence of large technology players from other

A. Kiselev (✉) · L. Silakova
ITMO University, Kronverksky Pr. 49, St. Petersburg 197101, Russia
e-mail: Arkadijk99@gmail.com

© The Author(s), under exclusive license to Springer Nature Switzerland AG 2023   45
A. Bencsik and A. Kulachinskaya (eds.), *Digital Transformation: What is the Company of Today?*, Lecture Notes in Networks and Systems 805,
https://doi.org/10.1007/978-3-031-46594-9_4

industries like Sber, Yandex, MTC which are replacing traditional media companies by using advanced technologies and large financial resources.

Taking into account the fact that the main source of income for media companies is advertising, which third-party companies buy for promotion [1], there is a dependence of the country's advertising market on external factors (foreign political situation, the state of the world economy) and internal factors (legislative regulation of the market, the emergence of new players from other industries).

Media companies and companies from other industries understand how important is digital transformation (DT) to adapt to the rapidly changing media environment and continue to satisfy audience needs. Successfully implemented digital transformation can help companies expand their business by creating new digital services and, as a result, business processes, improve customer experience and create new sources of income, new ways to monetize current products, enter new markets [2–5]. In the scientific field, the authors pay quite a lot of attention to the concept of digital transformation, however, often without reference to a specific industry [6–12]. Therefore, digital transformation in the media industry arouses particular interest for authors.

The goal of the digital transformation in media companies can be considered to increase the operational efficiency and obtain a synergistic effect by integrating various products and services into a single digital platform. The result of this process is the creation of a digital ecosystem.

There are many examples of companies that have built a digital ecosystem in the process of digital transformation. One of them is Microsoft, which creates and implements digital technologies both in third-party organizations and optimizes its own business processes and creates a full-fledged digital ecosystem. Amazon has created an ecosystem by bringing together their online retail, cloud services, marketplace, streaming services, and other products and services. As a result, company create a unique experience for customers and expand its business in various directions. Therefore, digital ecosystem is one of the possible results of the company's digital transformation. Company that has created a digital ecosystem can continue to carry out digital transformation based on the assessment of digital maturity and strategy, that they will choose.

According to a study by "Team A" (IT-consulting company), there are five steps to run a digital transformation business [13]. The figure below shows these stages, and we have also added comments at which stages of digital transformation the company is choosing a digital ecosystem as a result of digital transformation and make a detailed plan of strategy (Fig. 1).

Therefore, after digital maturity assessment, the company's management forms a strategic vision and decides how the company should change. At this stage, companies can choose the direction of creating a digital ecosystem. After DT management department was created, a DT strategy should be developed, in which it is necessary to develop a transformation plan with details of projects and initiatives.

Within the digital ecosystem, companies perform different roles based on their characteristics and competitive advantages. For example, various video streaming and audio streaming services tend to have a large user base, while usually always

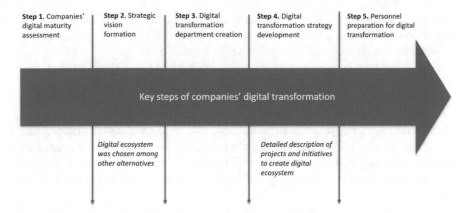

| Step 1. Companies' digital maturity assessment | Step 2. Strategic vision formation | Step 3. Digital transformation department creation | Step 4. Digital transformation strategy development | Step 5. Personnel preparation for digital transformation |

Key steps of companies' digital transformation

| | *Digital ecosystem was chosen among other alternatives* | | *Detailed description of projects and initiatives to create digital ecosystem* | |

Fig. 1 Key steps of companies' digital transformation

having low profitability [14]. It is beneficial for ecosystems to use a large user base of these services in order to offer them higher margin products within this network. The examples of high margin products can be real estate sales, e commerce [15].

Conceptually, an ecosystem is a network of companies, so companies need to change their business structure in order to become part of such a structure. In particular, digital ecosystems can be used to enable more efficient and flexible teamwork and information sharing across departments.

At the same time, the media ecosystem, according to a number of researchers, is a set of interdependent economic and technological components that form a certain integrity, form an interconnected structure of society, the unity of relations in the sphere of production, storage, distribution, exchange digital content in a single digital and interactive technological media environment [16–18].

An analysis of the Elibrary and Elsevier databases showed a lack of research on the development of media companies' ecosystems or the development of media services within the ecosystem. The reason for this may be that the largest business ecosystems in Russia include not only media assets, but also services from other industries (the most famous: e-commerce, banking, taxi, cloud services). For example, the Yandex company has several media assets (Yandex Music, Kinopoisk) that users can get through a Yandex Plus subscription. At the same time, thanks to the subscription, users receive an increased cashback in taxis, food delivery, e-commerce, which they can spend in the future within this ecosystem. In this regard, it is difficult to study only the media assets of this ecosystem, since the company's management can redistribute resources within the company for the sake of profit, so that public accounting and financial statements will differ from the actual data.

Despite the lack of research on this particular topic, there are works devoted to broader related topics. For example, there is a study that discover approaches to the regulation of ecosystems in Russia. Particular attention is paid to the role of the banking sector within the ecosystem, as data from banking transactions makes it possible to analyze customer preferences and adapt the ecosystem services to new customer preferences. However, the study compares ecosystems only by the type of platform solutions model within ecosystems (open, closed, hybrid), but does not consider the development of ecosystem media services [19].

In another study, ecosystems were considered as a tool for strategic development, but it remained unclear how to determine the most developed system [20].

In foreign sources, there are studies devoted to the study of the media ecosystems concept. The most fundamental is the study of the organization of economic cooperation and development, which describes the concepts of digital platforms and digital ecosystems [21]. However, this study is aimed at studying ecosystems and considering companies in all areas of activity. At the same time, the media segment in ecosystems is not considered separately, and ecosystems are not compared with each other.

There is also a study that aims to explore the ecosystems of digital content and services. Despite the narrowly focused title, the authors of the study consider the maturity of ecosystems not for specific companies, but for countries as a whole, therefore, the methodology implies the usage of criteria that are relevant for comparing the development of entire countries but not comparing the results of digital platforms/ ecosystems of specific companies [22].

In addition to research aimed at studying the digital ecosystem, there are works that explore the process of digital transformation in digital platforms and digital ecosystems. Thus, there is a foreign study that states that digital transformation can be expressed in the creation of a business based on a digital platform within the existing innovation ecosystem. At the same time, three levels are distinguished within this process: 1. The level of strategic decision-making and the level of management, 2. The co-innovation and coordination level, 3. the level of the joint production process [23]. In another work, they study the effects of digitalization in the context of digital ecosystems [24].

It should be noted that in Russia there is a study where the digital eco-system is considered as a means of digital transformation of the university [25]. However, digital transformation is a process that involves the usage of digital technologies at multiple levels of an organization. The same features apply to the digital ecosystem. However, to create a digital ecosystem, additional features are needed that may not be present in digital transformation. For digital transformation, there is no need to have one or more digital platforms, and also be present in several markets, unlike an ecosystem. But it is also possible to perform digital transformation after company become digital ecosystem. Therefore, the digital transformation process can be used both to create a digital ecosystem and to further develop the company, and then the digital ecosystem can become part of the digital transformation.

Therefore, the lack of relevant research on the development of media services among digital ecosystems makes this work relevant, given the fact that digital business transformation and digital ecosystems are currently popular areas of research.

The aim of the work is to develop a methodology for assessing the development of media segments of digital ecosystems in the media industry.

The tasks of the work are the following:

- Perform a review existing definitions of business ecosystems and digital ecosystems, and analyze ecosystem features
- Analysis of the general condition of the largest ecosystems in Russia and the world and their impact on the economy
- Analysis of the largest media companies in Russia: media services monetization model study, presence and ecosystem types, operating model
- Develop criteria for comparing ecosystems, as well as determine their weight for comparing media segments.

2 Methods and Materials

There are many companies in Russia that are actively developing digital ecosystems, while almost all companies have assets from different markets. The purpose of the work is to compare digital ecosystems by assessing media services development. Therefore, we analyzed those ecosystems that are at least partially present in the media market. 10 companies Yandex, Sberbank, VK, Megafon, MTS, Gazprom media, National Media Group (NMG), VGTRK, European Media Group (EMG), Russian Media Group (RMG) were taken for analysis.

The stages of the study were following:

- Analysis of the essence and features of the digital ecosystem. Existing research analysis and list of companies' definitions that need to be considered in the media market.
- Identified companies' analysis by type of core business and ecosystem, media assets presence and types, monetization and operational model of each company. An operating model is an abstract representation of the applied methods and procedures for implementing a corporate strategy in the daily activities of a company [26]. We used open sources to find information about each company. This stage will help to highlight the features of ecosystems in working with the media segment.
- Comparison of digital ecosystem by assessing media services development. For comparison groups of criteria will be created. In each group criteria will be defined and companies will be analyzed. The criteria aim to assess the development of the ecosystem as a whole, the media segment within the ecosystem, as well as media services within the ecosystem. Each criteria within the group has a weight, based on the importance and influence of this indicator. The data obtained for each criterion will be normalized and multiplied by weights to obtain the final value for the group of criteria. Further, the values for each group of criteria will be summed up for each company and the total values analyzed.

3   Results

Business ecosystems contribute to the development not only of themselves, but also of the countries in which the ecosystem operates. The development of digital platforms and ecosystems leads to country's economic growth, productivity and innovative activity growth in the economy, and the expansion of international trade. For example, in Western European countries, with an increase in the share of e-commerce in total retail market by 10 p.p., the annual GDP growth rate increased by 0.01 p.p. They also affect the labor market, inflation and prices of goods and services, and other macroeconomic factors [27].

In the scientific community, there are several variations of the ecosystem terms. The term business ecosystem was introduced by J. Moore–it is a flexible structure that includes people, firms interacting with each other to create and exchange values [28]. At the same time, there are similar concepts of the ecosystem that were given later and are more accurate and relevant to the current business. For example, an ecosystem is a network of interconnected companies that work together to achieve a common goal. One of the main goals of creating an ecosystem is to increase business efficiency by integrating various products and services into a single platform, which can lead to synergy and increased competitiveness. A similar definition is given by BCG: a business ecosystem is a dynamic group of largely independent economic players that create products or services that together constitute a coherent solution [29].

The concept of "digital ecosystem" is not formulated in countries legislation and is still rather informal. According to a study by the Gaidar Institute, a digital ecosystem is a broader concept in relation to a digital platform, while the main properties of ecosystems are similar to those of platforms. Digital ecosystems develop on the basis of digital platforms by connecting other digital services to them. Platform companies have the highest potential for becoming ecosystems [27].

According to researchers from the Gaidar Institute, the features of a digital ecosystem include:

- At least one digital platform among the company's services;
- Presence in more than two markets and/or industries;
- Integration of services with each other (both technical, for example, through a single technical solution, a common subscription, a loyalty program, a single ID or a super application, etc., and substantive–the benefits of obtaining several services for the consumer);
- High role of data about users and their actions, as well as ways of collecting, storing and processing them in the company's business model [27].

This list of features is the most accurate and applicable for determining digital maturity, while in order to compare companies in the future, the integration of services with each other and data collection will be further separated in order to assess the development of ecosystems.

There are several types of ecosystems: transaction ecosystems and decision ecosystems. Transaction ecosystems imply communication between producers of goods and services and customers through a single platform. Solution ecosystems imply the creation of a product by adding additional participants to the value chain that create additional goods and services [30]. A mixed model is also possible, which combines a solution ecosystem and a transaction ecosystem.

The building of business ecosystems and digital ecosystems takes place in order to improve the financial and operational performance of companies. In Russia and in the world, there are many companies from different industries that are actively developing as ecosystems to attract even more customers/users. Table 1 presents a list of the largest ecosystems in the world and Russia.

According to the table, only Meta has a core business related to media (social network), while every major ecosystem in the world has media assets available, but they are not the core business.

Currently, there are more than 10 digital ecosystems in Russia. Table 2 present information about the largest Russian ecosystems and their key businesses.

Based on the table above, we can conclude that there is no major ecosystem with media as a core business. At the same time, only about half of the companies have media assets. For comparison, below is a table with a list of the world's largest ecosystems.

To assess how ecosystem approach has influenced the results of these companies, it is necessary to assess the size of their historical capitalization. Figure 2 shows the dynamics of capitalization of the largest world and Russian ecosystems.

As you can see, the largest growth in capitalization was in American companies (Meta, Alphabet, Amazon), which form the ecosystem, while the cost of Chinese ecosystems is below than US companies, but over the past 10 years they have also grown significantly. Russian companies have not yet realized their maximum growth potential and this should be expected in the future (Fig. 3).

For comparison, the impact of Russian ecosystems on the global economy is shown below in the figure as a percentage of their share.

Table 1 List of the biggest global business ecosystem [27]

Company name	Core business	Media assets presence, yes/no
Meta	Social network	Yes
Alphabet	Search engine	Yes
Amazon	E-commerce	Yes
Apple	IT	Yes
Baidu	Search engine	Yes
Tencent	Gaming	Yes
Alibaba	E-commerce	Yes

Table 2 List of the biggest business ecosystem in Russia [27]

Company name	Core business	Media assets presence, yes/no
Sber	Banking	Yes
Yandex	Search engine	Yes
X5 Retail Group	Retail	No
Wildberries	E-commerce	No
Tinkoff	Bankinf	No
Ozon	E-commerce	No
VTB	Bankin	No
MTS	Telecommunication	Yes
Megafon	Telecommunication	Yes
Avito	Classified	No
VK (ex. Mail.ru)	Search engine	Yes
Gazprombank	Banking	Yes

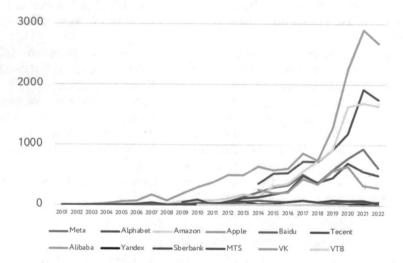

Fig. 2 Market capitalization of the largest global and Russian ecosystems, USD bn. [27]

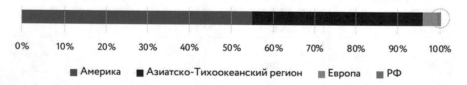

Fig. 3 Market capitalization distribution of the largest ecosystems by region, % [27]

Therefore, the share of Russian platform companies was only 0.76% in global capitalization, which indicates the low influence of Russian ecosystems on the global economy.

The authors agree with the opinion [31] that the main trends in the development of the Russian audiovisual market will remain consolidation, vertical integration and large digital ecosystems formation in partnership with traditional businesses (telecom, banks, etc.) with large subscriber bases.

To cover the entire media industry, it is necessary to consider media companies in traditional media segments (Radio, TV, press) and in digital segments (video streaming, audio streaming,). For this task, it is necessary to consider the largest representatives of each direction and analyze business and operational model, media services. Thus, the largest representatives of traditional media were identified in the context of each media segment. In the TV segment, there are three largest holdings (Gazprom-Media, National Media Group, VGTRK), in the radio segment, three largest companies are: the European Media Group, Gazprom-Radio (Gazprom-Media structure), Russian Media Group, and in the press segment Izvestia, Kommersant and Rossiyskaya gazeta are the largest newspapers [32–34]. However, the traditional and digital press was not considered further, since these assets cannot be attributed to ecosystems, since they do not have a digital platform and are presented either on one traditional market (press) or two traditional markets (press and radio). The following table describes media services, their business and operational models of the largest companies in their industry. Analysis of companies also include information about the presence/absence of ecosystems and their types (Table 3).

Based on the analyzed companies, we can conclude that transactional model prevails among the largest digital platforms, the mixed model prevails among TV holdings, and radio companies do not have their own ecosystem at all, since they do not have a digital platform.

Next, it is necessary to compare media segment of Russian ecosystems. To do this, it we need to determine the groups of criteria by which ecosystems will be analyzed. Criteria groups were compiled based on the fact that the most developed media segment in the ecosystem in Russia should have the following characteristics:

Media segment position in the company's ecosystem. To assess the significancy of media segment in ecosystem, we need to assess how many media services are present in this ecosystem and what share the media segment has in the revenue structure of media companies. Also, digital ecosystem development as a whole and the media segment is indicated by the number of digital services. These include products that only work online. Therefore, the share of digital services in the media segment should also be taken into account when analyzing the position of the media industry in the company's ecosystem.

Content quantity and quality. The main product of media company is content; therefore, we need to evaluate both its quantity and quality. Number of films/series/tracks/in the service and the number of active users per month (MAU) should be analyzed, because it reflects the content quality. Therefore, digital types of media services will be analyzed: video streaming, audio streaming, digital press.

Table 3 Media services description of the biggest digital ecosystem in Russia

Company name	List of media services	Ecosystem Type	Media assets monetization model	Operational model
Yandex	Yandex music–audio streaming Kinopoisk–video streaming Yandex Studio–content production	Transactional	Yandex Plus subscription (inc. media services) 300 rub./month. Advertising model is also working	Yandex Music and Kinopoisk buy license to use audio and video content from copyright holders and give access Yandex Plus subscribers to this content through the platform. Yandex also produce own video content
Sberbank	Sber zvook–audio streaming Okko–video streaming	Transactional	Sber prime subscription (inc. media services) 200 rub./month. Advertising model is also working	Sber zvook and Okko buy license to use audio and video content from copyright holders and give access Sberprime subscribers to this content through the platform. Okko also produce own video content
VK	VK Music–audio streaming	Transactional	Access to music and discounts on services from partners–169 rub./ month Advertising model is also working	VK buy license to use audio and video content from copyright holders and give access Sberprime subscribers to this content through the platform

(continued)

Table 3 (continued)

Company name	List of media services	Ecosystem Type	Media assets monetization model	Operational model
Megafon	Megafon music–audio streaming Start–video streaming Megafon TV–digital TV Megafon press–press aggregator	Transactional	Company is working by subscription model, and each service has separate price. Start–249 rub./month., Music–169 rub./month., Megafon TV connected with Start. Megafon Press–8 rub. / in month	Megafon Music works as a showcase of Sber Zvook and Yandex Music. Start produce and buy license to use video content from copyright holders and give access subscribers to this content through the platform. Megafon Press works as a showcase of Tele 2 press platform–kiozk
MTS	MTS Music–audiostreming Kion–videostreaming MTS Stroki–e-books MTS TV–Digital TV	Transactional	MTS Premium subscription 249 rub./month	MTS buy license to use audio and video content from copyright holders and give access s subscribers to this content through the platform. MTS has own content production MTS Originals. MTS provide digital TV
Gazprom media	Premier- Video streaming 9 TV channels 9 Radio stations 2 Newspapers	Mixed model	Premier subscription 299 rub./month	Company produce and distribute own content through Premier platform. Content from other producers can be also published on this platform
National media group (NMG)	More tv. – video streaming 5 TV channels 3 Newspapers	Mixed model	More tv subscription–299 rub./month. Company sell licenses on their own films and generate revenue from TV and press advertising model	Company produce and distribute video content. Content from other producers can be also published on this platform

(continued)

Table 3 (continued)

Company name	List of media services	Ecosystem Type	Media assets monetization model	Operational model
VGTRK	5 TV channels 5 Radiostations 1 Internet-resource	Mixed model	Company generate revenue from TV and radio advertising	Company produce and distribute own content through Premier platform
European media group (EMG)	9 radiostations	–	Company generate revenue from radio advertising	Company buy license to use audio content from copyright holders and give access to listeners
Russian media group (PMГ)	7 radio stations	–	Company generate revenue from radio advertising	Company buy license to use audio content from copyright holders and give access to listeners

- Ecosystem development in general. For assessing and comparing media segment development in ecosystem with other ecosystems, we need to assess ecosystems development for more objective comparison of media industries. The integration of ecosystems and the quality of the received client data will be compared to assess how ecosystem is developed. The ecosystems integration depends on the way the content is obtained. The most integrated type of ecosystem is the single subscription service. Integrations are also possible when a company has a digital platform that works on a subscription model, while there is another digital platform that is not connected to each other. Least integrated ecosystems have a digital platform that has assets from other markets/industries but no digital assets.

Based on the described characteristics of developed ecosystems, six groups of criteria were compiled for comparing companies:

- Media segment position in ecosystem (Media in general)
- Audio streaming
- Video streaming
- Digital press
- Integration
- Quality of received customer data

Criteria in each group have a certain weight, which will affect the calculation of the coefficient. Below is a table with a group of criteria and criteria that will be used to compare ecosystems, as well as their weight for calculating the coefficients (Table 4).

After compiling a group of criteria and criteria within each group, digital ecosystems with media assets were analyzed. The results for each company are presented

Table 4 List of criteria for assessing media services development in digital ecosystem

Group criteria	Criteria name	Criteria name (%)
Media in general	# media assets in ecosystem	33
Media in general	Media segment share in revenue structure	33
Media in general	% of digital assets in media segment	33
Audio streaming	Total number of songs, mln	50
Audio streaming	MAU (monthly active users)	50
Video streaming	Total number of films/series	50
Video streaming	Digital ecosystem has production center, 1–yes, 0–no	50
Digital press	MAU (monthly active users)	100
Digital ecosystem integration	1 point–maximum integration (single subscription on multiple services) 0, 5 point–there are a few digital platforms, but there is no direct synergy 0, 1 point–there is one digital platform and other services, which presented on different markets	100
Quality of received customer data	1 point–information source–banking transaction and search engine 0, 5 point–other information source	100

below. The information was obtained from open sources. Some indicators were calculated on the basis of the revealed information (Table 5).

Information about media assets and ecosystem as a whole were defined and analyzed. There are no data in some cells, because the ecosystem either does not have an asset in this part of the media, or information about the company was not found (for example, information about the number of tracks in radio stations).

The obtained data were normalized using the minimax approach to convert the results of each company to one scale and compare the values of each company with each other (Table 6).

Table 5 Media services development comparison in digital ecosystems in Russia

Company name	# media assets in ecosystem	Media segment share in revenue structure (%)	% of digital assets in media segment (%)	Total number of songs, mln	MAU music, mln	Total number of films/series, thousands	Production center presence	MAU press, mln	Integration	Quality of received customer data
Yandex	2	6	100	65	14	7,77	0	–	1	0,5
Sber	2	8	100	60	14	42	1	–	1	1
VK	1	7,9	100	400	41	–	0	–	1	0,5
MTS	4	0,7	75	65	1,5	10,1	1	–	1	1
Megafon	5	8	80	65	27,5	10	1	0,228	1	0,5
Gazprom-Media	37	100	27	–	40	4	1	12,6	0,5	1
NMG	20	100	80	–	–	40	1	47,2	0,1	0,5
VGTRK	12	100	17	–	30	7	1	19,6	0,5	0,5
EMG	9	100	11	–	36,4	–	0	1,4	0,1	0,5
RMG	7	100	29	–	35,1	–	0	–	0,1	0,5

Table 6 The final values of the digital ecosystems coefficient

Company name	Final values
Yandex	3,080,847,978
Sber	3,529,995,101
VK	2,854,042,298
MTS	2,852,303,599
Megafon	2,731,672,261
Gazprom Media	3,469,799,409
National Media Group	2,333,600,877
VGTRK	2,763,938,688
European Media Group	1,469,979,749
Russian Media Group	1,475,050,067

4 Conclusion

Based on analysis top five most developed media segment in ecosystems were defined: Sberbank, Gazprom-media, Yandex, VK, MTS. In most developed ecosystems, the media segment is not the main business and these companies generate revenue in other markets. The leaders are technology companies that have deep experience in various business areas and have a developed IT infrastructure that they use to improve their business.

Obtained results indicates that companies with media as a main business do not have developed ecosystems. The exception is Gazprom-Media, which is the largest media holding by TV coverage and ranks second by radio coverage in Russia. At the same time, the company has many digital assets [35] in various industries: Yappy (TikTok analogue), Rutube (Youtube an analogue) and many other lesser-known services. Gazprom-media has an advantage over other traditional media companies, because the parent company of GPM is Gazprombank, which collects data on customer transactions and can monitor how customer preferences change. Moreover, Gazprombank is developing its own subscription service "Ogon", which includes services from whole Gazprom holding (banking services, gas stations, media assets). Therefore, Gazprom Media is a part of general Gazprom subscription service.

There is a general trend among the most developed ecosystems,–the share of revenue from media is significant in the total revenue–up to 10%. This suggests that the most developed ecosystems use media assets not to make money on them. At the same time, media assets have a common advantage for the entire ecosystem, which is a wide coverage of the audience. The number of monthly active users of the largest audio streaming ecosystem services varies from 14 to 40 million for each platform. It means that significant part of the population uses digital media assets regularly. These services attract new customers and keep them within the overall ecosystem and allow this audience to interact with other company businesses that are the most marginal and generate more revenue.

The digital ecosystem, as a result of digital transformation, is a complex system. To maintain this system, it is necessary to have a sufficient amount of technological, material and labor resources and expertise to manage this form of business within the company. However, digital ecosystems appeared relatively recently and we should expect further growth in the number of digital ecosystems in Russia and the world.

At this moment, most of the largest traditional media companies in Russia do not have developed digital ecosystems, but focus on the development of their core business. There may be various reasons of this result: a lack of expertise and resources for the digital ecosystem creation or the unwillingness of the media holdings management to create such a form of business. Further research is planned to conduct surveys of companies in order to identify the main reasons of this problem, as well as suggest how media companies can carry out digital transformation to create a digital ecosystem.

References

1. Lambrecht A, Goldfarb A, Bonatti A et al (2014) How do firms make money selling digital goods online? Mark Lett 25:331–334. https://doi.org/10.1007/s11002-014-9310-5. Accessed 15 Apr 2023
2. Bolton RN, McColl-Kennedy JR, Cheung L, Gallan A, Orsingher C, Witell L, Zaki M (2018) Customer experience challenges: bringing together digital, physical and social realms. J Serv Manag 29, Article 5. https://doi.org/10.1108/josm-04-2018-0113. Accessed 22 May 2023
3. Masoud R, Basahel S (2023) The effects of digital transformation on firm performance: the role of customer experience and IT innovation. Digital 3:109–126. https://doi.org/10.3390/dig ital3020008. Accessed 05 June 2023
4. Kraus S, Durst S, Ferreira JJ, Veiga P, Kailer N, Weinmann A (2022) Digital transformation in business and management research: an overview of the current status quo. Int J Inf Manag 63:102466. ISSN 0268-4012. https://doi.org/10.1016/j.ijinfomgt.2021.102466. Accessed 17 Apr 2023
5. Silakova LV, Nikishina A (2021) Digital transformation of telecom providers management customer system: a process research and effects assessment. In: ACM International Conference Proceeding Series, pp 81–90
6. Vyugina DM (2016) Digital media business strategies in the context of changing media consumption. Mediascope, Issue 4. http://www.mediascope.ru/2233. Accessed 12 June 2023
7. RBC Trends (2023) How to distinguish digital transformation from digitalization. https://tre nds.rbc.ru/trends/industry/cmrm/606ae4c49a794754627d6161. Accessed 21 June 2023
8. Tsenzharik MK, Krylova YV, Steshenko VI (2020) Digital transformation of companies: strategic analysis, influence factors and models. Bulletin of St. Petersburg University. Economy, №3. https://cyberleninka.ru/article/n/tsifrovaya-transformatsiya-kompaniy-str ategicheskiy-analiz-faktory-vliyaniya-i-modeli. Accessed 15 May 2023
9. Mirzagayeva S, Aslanov H (2022) The process of digitalization of society: what does it lead to and what to expect in the future?. Metafizika (Journal) 5(4):10–21. eISSN 2617-751X. ISSN 2616-6879. OCLC 1117709579
10. Inozemtseva SA (2018) Technologies of digital transformation in Russia. In: Actual problems of economy, sociology and law, vol 1, pp 44–47
11. Popova SM (2019) On the issue of the concept of digital transformation of science. In: Trends and management, No 4, pp 1–16. https://doi.org/10.7256/2454-0730.2019.4.31941. https://nbp ublish.com/library_read_article.php?id=31941. Accessed 02 June 2023

12. Selina MV (2021) Digital transformation changes in the economy and the social sphere under the influence of technology. In: Digital transformation of industries: starting conditions and priorities. XXII April international scientific conference, jointly organized by the National Research University Higher School of Economics and Sberbank
13. KMDA (2020) Digital transformation in Russia 2020 overview and recipes for success. https://drive.google.com/file/d/1xVK4lSanDZSCN6kGAHXikrGoKgpVlcwN/view. Accessed 15 Mar 2023
14. O'Leary L (2023) Why it's been such a brutal summer for streaming. https://slate.com/techno logy/2022/08/streaming-trouble-hbo-max-netflix.html. Accessed 10 June 2023
15. GCTV Staff (2022) Top 4 most profitable industries in 2022. https://gctv.com/most-profitable-industries-2022/. Accessed 14 May 2023
16. Russian media industry: digital future. M.: MediaMir, (2017)—160. Accessed 17 May 2023
17. De Prato G, Sanz E, Simon JP (2014) Digital media worlds: the new economy of media. Palgrave MacMillan, London
18. Veselova SV (2016) Russian advertising yearbook. M.: Ros. acad. advertising
19. Ecosystems: approaches to regulation. Report for public consultations. Report of the Bank of Russia. http://www.cbr.ru/content/document/file/119960/consultation_paper_020 42021.pdf. Accessed 25 May 2023
20. Ivankova GV, Mochalina EV, Dubolazova YA (2023) Digital ecosystem: trend of strategic development of Russian companies. P-Economy. T. 16, №1, C. 7– 20. https://cyberleninka.ru/article/n/digital-ecosystem-trend-in-strategic-development-of-rus sian-companies/viewer. Accessed 04 June 2023
21. OECD (2019) An introduction to online platforms and their role in the digital transformation. OECD Publishing, Paris. https://doi.org/10.1787/53e5f593-en. Accessed 07 May 2023
22. Strategy &. Understanding digital content and services ecosystems. The role of content and services in boosting Internet adoption. https://www.strategyand.pwc.com/m1/en/reports/und erstanding-digital-content-and-services-ecosystems.pdf. Accessed 20 Apr 2023
23. Valkokari K, Hemilä J, Kääriäinen J (2022) Digital transformation—Cocreating a platform-based business within an innovation ecosystem. Int J Innov Manag 26(3). https://doi.org/10. 1142/S1363919622400163. Accessed 01 June 2023
24. Marton A (2022) Steps toward a digital ecology. Ecological principles for the study of digital ecosystems. J Inf Technol. https://doi.org/10.1177/02683962211043222. Accessed 19 Mar 2023
25. Neborsky EV (2021) Digital ecosystem as a means of university digital transformation. World of Science. Pedagogy and psychology. №4. https://cyberleninka.ru/article/n/tsifrovaya-ekosis tema kak sredstvo tsifrovoy transformatsii universiteta. Accessed 23 Mar 2023
26. Ross JW, Weill P, Robertson DC (2006) Enterprise architecture as strategy: creating a foundation for business execution. Harvard Business School Press, Boston. Accessed 29 May 2023
27. Digital ecosystems in Russia: evolution, typology, approaches to regulation. Institute of Economic Policy named after ET Gaidar. https://www.iep.ru/files/news/Issledovanie_jekosi stem_Otchet.pdf. Accessed 12 Mar 2023
28. Moore JF (1993) Predators and prey-A new ecology of competition. Harv Bus Rev 71(3):75–86
29. Ulrich P, Reeves M, Schüssler M (2023) Do you need a business ecosystem? https://www.bcg. com/publications/2019/do-you-need-business-ecosystem. Accessed 21 May 2023
30. Heads&Hands (2023) Is it worth creating a business ecosystem: consider the advantages and disadvantages. https://vc.ru/services/121003-stoit-li-sozdavat-biznes-ekosistemu-rassmotrim-preimushchestva-i-nedostatki? Accessed 16 May 2023
31. Evmenov AD, Blagova IY (2020) Russian market of digital audiovisual content: analysis and development forecast. Petersb Econ J №2. https://cyberleninka.ru/article/n/rossiyskiy-rynok-tsifrovogo-audiovizualnogo-kontenta-analiz-i-prognoz-razvitiya. Accessed 29 May 2023
32. RU-1000: The largest companies in Russia in 2022. https://www.oborudunion.ru/ratings/ 33222. Accessed 04 June 2023

33. Sergeevna KD (2023) Who owns the major radio stations in Russia: what do the extracts from the Unified State Register of Legal Entities say. https://moneymakerfactory.ru/spravochnik/kru pneyshie-radiostantsii-rossii-za-2018/. Accessed 21 May 2023

34. Medialogy. Rating of newspapers, magazines, TV channels, Internet resources. https://www. mlg.ru/ratings/. Accessed 25 Mar 2023

35. Gazprom media–Homepage. https://www.gazprom-media.com/ru. Accessed 15 June 2023

Assessing the Product Maturity of the IT Team in the Context of Digital Transformation

A. V. Ivanov◉ and L. V. Silakova◉

Abstract In a changing market, it is important for a company or a specific team not only to create a product, but also to be able to adapt to rapid changes in the market or customer requirements. Such adaptability requires constant changes in processes and approaches. One of the key market demands is the digital transformation of company processes. The transition to digital transformation requires tools for assessing product maturity that can provide information for a multifactorial analysis of the current product approach. The chapter discusses and analyzes the well-known models for assessing the maturity of OPM3, PMM, PMMM, IMPULS Industry 4.0 Readiness. The authors propose the introduction of a new methodology for assessing the product maturity of an individual team, and also give an example of evaluating real-life teams and conclude about the advantages of such an approach for a detailed analysis of the work of product teams.

Keywords Project management · Maturity in project management · Digital transformation · Technology assessment · Maturity model · Comparison of project management maturity models · Industry 4.0 · Big data

1 Introduction

The modern information technology market is characterized by rapid development and high competition. In such conditions, for the successful work of the team in the field of automation of information technologies and telecommunications, it becomes important not only to create high-quality products, but also to constantly increase the level of product maturity of their teams. According to the PM Solution Project Management Institute, about 37% of project failures are caused by a lack of clear goals [1]. To achieve the clearly set goals, companies need to increase the level of

A. V. Ivanov (✉) · L. V. Silakova
Faculty of Technological Management and Innovations, ITMO University, Saint-Petersburg, Russia
e-mail: artemi.01@mail.ru

© The Author(s), under exclusive license to Springer Nature Switzerland AG 2023
A. Bencsik and A. Kulachinskaya (eds.), *Digital Transformation: What is the Company of Today?*, Lecture Notes in Networks and Systems 805,
https://doi.org/10.1007/978-3-031-46594-9_5

maturity of their product. According to Kerzner, companies may need up to 7 years to reach maturity in project and process management [2]. However, there are many small businesses and start-ups that work with agile and agile methodologies that approach business systematically and methodically from the start. However, the question remains of maintaining this methodical approach to scaling, since well-known evaluation tools are not always capable of scaling or are capable, but cannot reliably and objectively reflect the state of affairs in the analysis of multifactor management of projects or product teams. Using the popular SAFe and DAD methodologies, we can talk about the importance of scaling fast IT projects [3]. Key factors of such strategy are not only in the size of the team or the company. The strategy includes taking into consideration geographical, political conditions as well as the level of technological advancement, legislative compliance with local authorities and the complexity of both customer and business approach. Each team has to adjust the strategy according to the specific conditions of such business environment. This is where the need arises to evaluate the effectiveness of such a strategy and search for weaknesses and strengths of the team for further improvement. Assessing weaknesses and strengths, and deeply transforming internal and customer processes are important elements of digital transformation. For instance, a team of five people working in one office, which is engaged in a regulated project, will act differently compared to a team of 40 people distributed across several offices and carrying out an unregulated project [4, 5]. The improvement and transformation of processes is inextricably linked with the introduction of innovations, it is because of such a request from the market that companies begin to pay special attention to innovations. An excellent practice in this regard is to delegate the functions of innovative business analysis to internal departments that test new solutions and, based on the data obtained, conclude that it is advisable to implement them at a higher level within the company.

The main purpose of the chapter is to evaluate product teams based on the proposed methodology, as well as to determine the possibility of its application on the example of individual projects, and not at the level of the entire operational activity of the company.

2   Definition of Concepts

2.1   *Product Maturity*

To formalize the requirements and criteria during the formation of the method of assessing product maturity, it is necessary to understand what product maturity is. The maturity of the team's product is the group's compliance with two blocks of criteria, soft and hard [6]. The soft block implies compliance of the work with the requirements, priorities and values of the enterprise in the broad sense, preservation and adherence to the established corporate culture. Rather tough block characterizes the team's compliance with external market conditions in terms of technological compatibility of equipment and technologies, the team's skill level and, consequently, in terms of flexibility of work to create a product for the client. A product maturity assessment model is a list of criteria that can be used to assess the current state of the company. There are several models that formulate certain stages of maturity of the team and the company as a whole [7]. As organizations move toward digital transformation, IT staff are often at the forefront of these efforts. The success of any digital transformation initiative relies heavily on the maturity of the IT staff. In this chapter, we look at the concept of product maturity in the context of the IT team and discuss how to assess it in the context of digital transformation.

Digital transformation is a term that seems extremely incomprehensible to many. There are still discussions regarding both the definition of digital transformation itself and the list of criteria, the presence or absence of which could accurately indicate the conformity of this very transformation.

Anyway, many experts agree that digital transformation is a logical continuation of the industrial and digital revolutions that have taken place in the world. To form an understanding, we propose to consider digital processes in the following form. The first stage of changing and adapting business processes was digitization, which was dictated by the need to duplicate documents electronically to increase the speed of internal processes. After digitization, there was a need and a desire to speed up the processes more through digitalization. Digitalization in this way is the embedding of digitized and digital documents into daily business processes and operational activities. Such innovations have been strengthened due to the transition of many public services to similar solutions to simplify the work of citizens who act as consumers of state services.

At the moment, digitalization of processes remains the dominant solution, but now, as before, there is a desire and need in competitive conditions to further increase the speed and improve the quality of decisions made with the help of increasingly popular machine learning and artificial intelligence tools.

The pioneers in the field of digital transformation have become the largest international technology corporations, which have begun to adjust their positioning strategy in the market. The main and key elements and value of digital transformation in the current understanding are embedding the results of digitalization into independent and independent functioning chains of internal processes. The trend in the development of this approach is the creation of automated business processes that require less and less "manual" decision-making. A well-described example is the creation of digital twins that aggregate large amounts of data and analyze them in a mode close to real-time mode. At the same time, it is difficult to overestimate the prospects that this may open in the near future, thanks to the constant improvement of existing tools for storing, processing and interpreting of Big Data.

Product maturity refers to the degree of complexity and completeness of a product or service. In the context of an IT team, product maturity can be thought of as the level of skill and effectiveness in delivering technology solutions. A mature IT team is a team that can consistently deliver high-quality technology solutions that meet the needs of the organization and its customers. Assessing the product maturity of the IT staff is critical in the context of digital transformation. Digital transformation initiatives often involve implementing new technology, changing business processes, and rethinking how the organization operates. A mature IT team can effectively support these initiatives and help the organization achieve its digital transformation goals. We may look at some key factors in assessing the product maturity level down below.

1. Technology stack

This item refers to the presence in the company and in a single team of a pool of technological solutions that they use to solve emerging problems. In the context of digital transformation, it is important that such solutions correspond to the company's strategy, and employees and team members understand their value and practical significance. The qualification of specialists, which allows implementing such solutions, is also worth noticing here, because the company should act in all the ways and parts of its business accordingly.

2. Processes and methodologies

A mature IT team should have clearly and competently structured processes that imply a specific project methodology. Business processes should eliminate uncertainties and reduce the human factor as they become more mature.

It is important to note that project methodologies in their pure form are rarely suitable for effective business in practice. That is why it is important to use the best practices of other companies and benchmarking similar solutions in this area. Examples of adaptation and synthesis of several methodologies can be called Scrumban, from the name of which it is clear that it uses Kanban and Scrum tools at once during project development. Again, it is important to meet the requirements of digital transformation, which imply resistance to change and rapid adaptation. Processes and methodologies should be implemented not only at the company's management level, but should also be understandable and transparent to all project specialists.

3. Skills and expertise

The logical continuation of the technological stack and optimized processes is worth noting skills and expertise. Here we are talking not only about the availability of competencies, but also their continuous improvement in accordance with market demands. The ability to approach problems outside the box is often called a soft skill, but it is inextricably linked with hard skills. Professional and supra-professional skills in a rapidly changing business climate certainly correspond to the goals of both digitalization and digital transformation with an improvement in the quality of decisions made and with an improvement in overall efficiency.

4. Customer orientation

Customer orientation is another important element of digital transformation, which often involves changing the approach to communicating with customers and building new channels for delivering product value. A particularly important element of working with the client is feedback, which, with the development of artificial

Table 1 Publications selected during the literature review process

Paper	Year	Authors
A review of Industry 4.0 Maturity Models: adoption of SMEs in the manufacturing and logistics sectors [8]	2023	Hussein Magdy Elhusseiny, José Crispim
Digital maturity model for research and development organization with the aspect of sustainability [9]	2023	Krzysztof Jacek Kupilas, Vicente Rodriguez Montequin, Javier García González, Guillermo Alonso Iglesias
Project Management Maturity Models: proposal of a framework for models comparison [7]	2023	Léa Domingues, Pedro Ribeiro
Investigation on the acceptance of an Industry 4.0 maturity model and improvement possibilities [10]	2022	Alexander Kieroth, Manuel Brunner, Nadine Bachmann, Herbert Jodlbauer, Wolfgang Kurz
Information security management maturity models [11]	2022	Natalia Miloslavskaya and Svetlana Tolstaya
A pilot study: An assessment of manufacturing SMEs using a new Industry 4.0 maturity model for manufacturing small- and middlesized enterprises (I4MMSME) [12]	2022	František Simetinger, Josef Basl
Maturity model for collaborative R&D university-industry sustainable partnerships [13]	2021	Cláudia Silva, Pedro Ribeiro, Eduardo B Pinto, Paula Monteiro

intelligence and analysis tools, allows you to create predictive models and see cause-and-effect relationships within most areas of the company's activities.

In conclusion, it can be noted that the assessment of the maturity of the product team carries great benefits for both sides of the interaction. The business receives data that allows it to understand and decompose the elements of its effectiveness, and the consumer, in turn, receives a high-quality product that takes into account his own wishes and comments as an element of the market. Such a tandem fully corresponds to the goals and values of digital transformation and improves the quality of business decisions.

In the course of the work, it was possible to analyze a number of articles (Table 1) to form evaluation criteria.

As a result of studying existing solutions, a pool of current maturity models was selected for evaluating product teams (Table 2).

2.2 OPM3 (Organizational Project Management Maturity Model)

The maturity model of organizational project management was developed in 2003 by the Institute of Project Management. This model differs from others in that, like PMBoK, it is, in fact, a set of best practices and sets of instructions for using them to

Table 2 Maturity models selected during the literature review process

Model	Maturity levels	Dimensions
OPM3 Organizational Project Management Maturity Model	*No system*	*Multi-dimensional* *Progressive stages*: Standardization Measurement Control Continuous Improvement *Three domains*: Project management Program management Portfolio management [14] *Life-cycle*: Initiating Planning Executing Controlling Closing
KPMMM Kerzner Project Management Maturity Model	*5 levels* Common Language Common Processes Singular methodology Benchmarking Continuous Improvement	*7 dimensions* Terminology Adaption Methodology Planning Benchmarking Improvement Risk analysis
PPMMM Prado Project Management Maturity Model	*5 levels* Initial Known Standardized Managed Optimized	*7 dimensions* Competence in Project and Program Management Competence in Technical and Contextual Aspects Behavioral Competence Methodology usage Computerization Usage of the convenient Organizational Structure Strategic Alignment
IMPULS Industry 4.0 Readiness	*6 levels* Outsider Beginner Intermediate Experienced Expert Top performer	*6 dimensions* Strategy and organization Smart factory Smart operations Smart products Data-driven services Employees

conduct a maturity assessment. The model is a multidimensional structure in which various parameters can be evaluated, including both operational efficiency and a comprehensive assessment of individual stages of interest to us [15]. Organizational Project Management Maturity Model (OPM3) Cycle includes 3 main stages:

1. Knowledge

Step 1. Preparation for evaluation: A stage that includes the formation of the content of the model, criteria, parameters, as well as the choice of evaluation methods.

2. Assessment

Step 2. Evaluation: A stage that includes the evaluation process itself according to developed, defined and agreed criteria. At the same stage, the analysis of existing practices is carried out both within the company and in the foreign market. As a result of this step, it is necessary to conduct a thorough investigation at a sufficient level to formulate plans for improvement.

3. Improvement

Step 3. Improvement Plan: At this stage, the improvement plan formulated in step 2 is finalized with prioritization of the desired improvements and results and the development of a roadmap of work.

Step 4. Implementation of improvements: At this stage, the proposed changes are introduced into the work processes to improve work efficiency and increase the level of maturity.

The cycle must be repeated continuously to achieve or maintain the desired level of maturity.

2.3 KPMMM (Kerzner Project Management Maturity Model)

KPMMM is a project management maturity model developed by Harold Kerzner, a renowned project management expert. The model aims to improve project management practices in an organization by assessing maturity levels in five key areas: strategic alignment, project management methodology, Project Performance Indicators, project management culture, and human resources. The model provides organizations with a plan to increase the maturity of project management by identifying weaknesses and implementing best practices. The Kerzner Project Management Maturity model offers 5 maturity levels, each marked by certain criteria. There are general frames of the model:

1. Each level following the other cannot be started until the previous one is completed;
2. Levels may overlap;
3. Risks may arise at each maturity level [16];
4. There are traps or obstacles that prevent you from reaching the next level;

5. Different levels of maturity of the project management system at the enterprise are characterized by a number of problems related to the resistance to innovation of employees and departments.

Each of the five levels represents a different degree of maturity of project management.

1. Common Language (Level 1)–this level assumes that the company implements universal project management tools and project teams can "communicate in the same language".
2. Common Processes (Level 2) involves not only understanding each other, but also integrating the company's processes with each other, in which a common process is developed with the possibility of scaling and repetition.
3. Singular methodology (Level 3) involves the standardization of the work of various teams according to a single methodology, which the company defines for itself, in order to achieve a synergistic effect.
4. Benchmarking (Level 4) implies the definition of parameters for benchmarking and its constant implementation.
5. Continuous Improvement (Level 5) - there is a continuous improvement of processes within the company based on benchmarking data.

Separately, it should be noted that the model assumes the presence of a risk matrix, in which the level of Singular methodology carries the greatest risk, since companies are usually flexible in choosing the methodology of their work.

2.4 PPMMM (Prado Project Management Maturity Model)

PPMMM is a maturity model that considers various management states, from immature project management methods to integrated and optimized systems. The Prado project management maturity model describes 5 maturity levels.

1. Original (Level 1)–At this mature stage, the company, unlike many other models, does not use project management tools and does not have standardized procedures in its work.
2. Known (Level 2)–at this level, the organization begins to implement standard project management processes and procedures. The main goal is to increase the efficiency and predictability of the project.
3. Standardization (Level 3)–At this level, the organization creates project management processes and procedures that are fully integrated into the organization's business processes. Its main goal is to improve the quality and reliability of products, as well as reduce the time and cost of their development.
4. Escrow (Level 4)–At this level, the organization actively uses methods and tools for quality and process management. The main goal is to constantly improve the process and increase the efficiency of the team.

5. Optimization (Level 5)–At this level, the organization reaches the highest maturity in project management. Its main goal is the continuous improvement of technological processes and innovative methods of product development.

2.5   IMPULS Industry 4.0 Readiness

IMPULS Industry 4.0 Readiness is an assessment methodology focusing on the readiness of a company or corporation to transition to the era of Industry 4.0. The essence of the assessment is to match the strategy with current trends like Artificial Intelligence, multi–level automation, the Internet of Things, etc. The proposed methodology includes 6 aspects of evaluation: strategy and organization, "smart factory", "smart operation", "smart products", data-based services and employees. Smart factory is an approach to the organization of production of goods or services. Within the framework of our topic, we are talking about an IT product and the projects inside its activity. Smart operation is the general name of a list of approaches and tools for automating and debugging business processes from order receipt to delivery to the end user. Smart products reflect the decision-making about the characteristics of the manufactured product based on the collected data on demand and feedback on completed contracts. Continuous improvement and adaptation to new market conditions of demand is another key factor here for the transition to industry 4.0. Data-based services and employees, respectively, summarize the necessary criteria for optimizing external communication processes with consumers, as well as internal processes of employee interaction at the level of a single and at the level of interaction of full-fledged cross-functional teams.

Each block meets the current requirements and challenges of Industry 4.0. and is filled with constituent criteria. Based on expert evaluation, one of six maturity levels is determined–from Outsider (Level 0), when the company does not use project management tools, to Maximum productivity (Level 5), when the company constantly evaluates and improves its own processes.

2.6   Maturity Models in Terms of IT, Automation and Telecommunications

The considered models offer excellent options for assessing the maturity of the company and processes, but they often seem insufficiently flexible to evaluate the work of individual teams and identify point weaknesses in the work of the product team [17]. This suggests the need to introduce such a tool that does not conflict with the maturity assessment models analyzed. Product maturity assessment is an important tool for companies to understand how successful their new products are and how well adjusted the internal processes are. It includes analysis of various factors, such as product quality, competitiveness, adaptability, etc. Product maturity assessment can help companies identify problems and shortcomings in their products and what improvements need to be made to improve their quality and competitiveness.

Separately, it is important to note the role of Machine Learning (ML) and Artificial Intelligence (AI) in business activities. Companies are increasingly trying to use big data analysis tools, including to understand the client's path, to work with feedback and a number of other important positions for a particular business.

Modern project management tools also try to aggregate information from the public field, for example, from the Internet, social networks, various thematic forums, etc.

Although using AI and ML to assess the maturity of a product has many advantages, it also has a number of disadvantages. The most pressing issue at the moment is the quality of data interpretation. Taking into account the current development of technology, there are great advances in the quality of such an interpretation in automatic mode, but do not forget that with the growth of the company, the price of error increases with incorrect interpretation or interpretation that does not take into account the specifics of the industry or an individual company.

The described analytics tools are actively used in business analytics, but there is still no widespread introduction of the so-called innovative business analysis. This type of business analysis involves the transfer of part of the functions of testing and evaluating the prospects for the introduction of new solutions to individual departments and qualified employees who pursue the goal of increasing the efficiency of the company or improving the quality of manufactured or developed products. The need for such a solution, as one might guess, was dictated by the fact that many enterprises, often operating within a large group of companies, are not interested in introducing innovations in practice, since it takes time and carries not only benefits but also risks. It is in this case that delegating the relevant tasks to business analysts working within the entire perimeter of the group of companies and having a different view on process optimization is appropriate.

Maturity assessment is inextricably linked with project management and is its logical component. The use of an innovative approach and the introduction of machine learning and AI tools really opens up great prospects, but the issue of quality control of output data remains important, taking into account historical data and geographical, social, political, economic or other specifics of doing business.

As a result of the analysis, given that we did not have a goal to determine the best of the existing evaluation models, but in the context of IT [9], automation and telecommunications sector under consideration, it is worth noting that the PPMMM (Prado Project Management Maturity Model) model is the most flexible model for working with companies and their projects in this area. A clear advantage is the high versatility of such a model [18, 19], but there is also a clear disadvantage compared to OPM3, since the Prado model does not involve multi-dimensional measurements and does not include, for instance, the Portfolio management evaluation [14].

3 Methods and Methodology

Modern product and project teams in the field of automation of information technologies and telecommunications work in a short time and with constantly changing requirements. Various flexible methodologies can be used to achieve the set goals and timely completion of tasks. At the same time, not classical methodologies such as Scrum, Lean, Kanban, but their adaptations and hybrids are becoming increasingly popular. These include such methodologies as Scrumban, Scaled Agile Framework (SAFe), Disciplined Agile Delivery (DAD) and others [4].

Scrumban is a hybrid flexible methodology that combines elements of Scrum and Kanban at once. Scrumban uses an iterative and step-by-step Scrum process and combines it with Kanban elements. This hybrid is well suited for organizations seeking a smooth transition to Kanban. The process begins with using the Scrum board, but after completion of the work it is transferred to the Kanban board to visualize the current processes and evaluate them from a different point of view.

Scaled Agile Framework (SAFe) is a semi-popular flexible methodology designed for large enterprises. An example of companies using SAFe is Sberbank, Gazpromneft and others. Such a comprehensive methodology provides the foundation for operational organizational changes. In essence, these are recommendations for scaling or implementing Agile, and therefore such a framework is often used by organizations to achieve consistency between several teams, which can be assumed from a sample of large companies. Within this framework, the focus is on the availability of flexible teams, releases and flexible portfolios. The importance of creating effective cross-functional teams and their consistency forms the basis of the framework to achieve high quality products.

Disciplined Agile Delivery (DAD) goes beyond the well-known Scrum and Kanban methodologies. It offers a lean and flexible approach to project management. This approach provides a comprehensive view of the entire product lifecycle, from design to decommissioning. Also, within this approach, recommendations are provided for adapting flexible practices to the unique needs of each existing project. A distinctive feature of DAD is the absorption of the best design practices with the selection of a pool of suitable solutions, which at the same time do not contradict the values of Agile and DAD.

Each of these flexible structures has its own strengths and weaknesses. Scrumban provides greater flexibility in job management, SAFe offers a comprehensive flexible scaling platform, and DAD offers a holistic view of the product lifecycle. Organizations should choose the flexible platform that best suits their unique needs and work environment.

When choosing, it is important to take into account factors such as the size of the organization, the level of complexity of projects and the number of teams within the company. It is also important to take into account the organizational culture and the level of support for flexible practices at all levels of the organization up to senior management.

With the right flexible structure, organizations can deliver high-quality products and services in a flexible and scalable manner.

The described methods are designed to achieve the set goals in a short time while minimizing costs. The use of project methodologies without analyzing their effectiveness is often, but not always, the most effective solution. Of course, as in many other industries, the thoughtfulness of all business processes and projects reflects the maturity of the company, but this requires additional time and intellectual costs, which ultimately result in money. In this chapter, it is proposed to understand the decomposition of productivity as the division of the results of the team's work into key blocks that reflect the interests and development strategies of the company. This decomposition is very valuable because it not only reflects key performance indicators, but also allows you to identify specific reasons for achieving or not achieving results. In the first case, the information obtained allows us to implement best practices in the next iteration, while in the second case, conclusions can be drawn to overcome errors and prevent performance degradation in the future. Considering the promotion strategy, all this data is inherently specific to the company and can be used to prepare models for machine learning and create more accurate predictive models and digital doubles, which corresponds to the digital transformation strategy.

Digital transformation is not just the integration of digital products into companies. The integration of digital solutions into business processes is also an important prerequisite for meeting the requirements of digital transformation [20]. The creation of digital doubles can be defined as the next level of digitalization of processes. All these steps are aimed at creating a highly adaptive and flexible company that improves the quality of its services and minimizes costs. Since the creation of digital solutions and the digital transformation of companies of various maturity levels is, in fact, a project, proper project management is required. At the management level, it is important not only to understand whether there is a result of the work, but also to understand the internal maturity of the team. Taking into account the analysis of existing maturity models and current project management methodologies, there is a need to evaluate not only the entire company's activities as a whole, but also a more focused assessment of specific product teams around the perimeter of the company. Interest in this kind of assessment arises not only among companies that focus on the foreign market, but also among corporations that create their own products and services [4].

Within the framework of the topic under consideration, it is important to see the full picture of the world and assess the prospects for the implementation of assessment methodologies. It is possible that the assessment of product maturity will be less in demand by small startups with their own financing. However, in recent years, the culture of student entrepreneurship has been actively developing in different countries of the world. A good example in the proposed context are universities with their own accelerators. Startups implemented in whole or in part at the expense of the university endowment can implement maturity assessment tools and at the same time develop the general level of culture of IT specialists on the way to global digital transformation.

In turn, the implemented tools will allow the university to better monitor the effectiveness of investments and the direction of its own development in both the entrepreneurial and educational spheres. Student entrepreneurship and, in particular, student start-ups play an important role in the global digital transformation. Students from and related fields enter the labor market after graduation and hold positions in leading technology companies. At the same time, it is important that the knowledge and acquired skills meet the requirements and challenges of the time.

As already mentioned, digital transformation includes a number of important requirements for the transition to a fundamentally new business model that takes into account the creation of a digital infrastructure for integration into current non-automated processes, as well as the creation of an ecosystem that forms an environment for innovation and the training and development of specialists with relevant competencies, including both professional skills and soft skills on communication and interaction.

The question of the necessity and expediency of implementing evaluation tools in teams developing hardware products remains open. By analogy with the complexity of implementing the Agile approach in such teams, the assessment of product maturity at the current stage of the fragmentation and informalization of many product requirements remains a field for discussion and a direction for the development of appropriate methodologies.

For such an assessment, the following methodology is proposed (Table 3), which will reflect both the goals set by the company's management and the maturity of the work of the team itself.

In accordance with the assessment of the team's product maturity [21], it can be attributed to one of the maturity categories in accordance with the PPMMM (Prado Project Management Maturity Model) model discussed earlier. It is proposed to set the boundaries in accordance with the proportional distribution between 100 points and 5 levels.

At the stage of summing up the Block points, it is proposed to draw a conclusion about the real weight of the block based on the fulfillment of the key criteria that are included in it. The following is a general formula for evaluating block scores.

Calculation of Block X points:

$$m_X = \frac{m_{X1} + m_{X2} + \cdots + m_{Xn}}{100} \cdot M_X \tag{1}$$

where:

M_X–*maximum block weight, points*

m_{Xn}–*criterion weight, points*

m_X–*real block weight, points*

Next, it is proposed to consider the assessment of the maturity of the Technology block (Table 4) selected at step 1 of determining the company's values.

Table 3 Methodology

Step	Stage	Example and comments
1	Determining the value pool of the product team and its ranking	Documentation > Product or Documentation < Product, etc
2	Allocation of key Blocks (in accordance with an optimization strategy)	Documentation, Architecture, Technology, Testing, Tracking, Support, etc
3	Assigning each Block the corresponding maximum weight in points	Weight assignment is carried out on the basis of an expert assessment in accordance with the value of the Block for the product team *(The sum of all points should be 100)*
4	Filling the Block with evaluation criteria	Technologies: (1) The source files are in a single repository (2) The code text is written in a single style and in accordance with current requirements
5	Assigning each criterion an appropriate weight in points	Weight assignment is carried out on the basis of expert evaluation in accordance with the value of the Block for the product team *(The sum of all points within one Block should be 100)*
6	Conclusion on the fulfillment of the criteria	Based on this assessment, a conclusion is made about the criteria in the form of fulfilled/ not fulfilled *(If the criterion is partially fulfilled, it is concluded that it is not fulfilled)*
7	Summing up Block scores	The Block scores are equal to a fraction of the maximum weight *(a percentage equal to the sum of the points for the criteria within the Block)*

Similarly, we can evaluate the remaining selected blocks and make a conclusion about the maturity of the team. Assume that the remaining Blocks weigh proportionally to the Technology Block. Then we get that the total score will relate to 100 in the same way as 11 points relate to 20. As a result of the evaluation, the product maturity in the selected case will be 55 points out of 100. According to the distribution of points and maturity levels in the PPMMM (Prado Project Management Maturity Model) model (Table 5).

We get that such a team corresponds to the 3rd level of maturity PPMMM Standardized (Level 3)–at this level, the organization has established project management processes and procedures that are fully integrated into the business processes of the organization.

The advantage of the proposed methodology is the adaptability of the evaluation criteria in accordance with the company's priorities [21]. Such a methodology for evaluating projects and product teams is a convenient tool for management accounting and internal analysis, and is also essentially a reflection of the company's internal process approaches and automatically changes at the 1st stage of determining the value pool [22]. A separate vector for further development of the maturity model may be the assessment of time costs. Such an assessment can

Table 4 Example of evaluation

Step	Stage	Example
2	Selecting a Technology Block (one of the highlighted ones)	**Technologies**
3	Assigning the Block the appropriate maximum weight in points	Technologies (20 points)
4	Filling the Block with evaluation criteria	**(1) The source files are in a single repository** **(2) The code text is written in the same style** **(3) The text of the code meets the current requirements**
5	Assigning each criterion an appropriate weight in points	(1) The files are in a single repository **(25 points)** (2) The code text is written in the same style **(45 points)** (3) The text of the code meets the current requirements **(30 points)**
6	Conclusion on the fulfillment of the criteria	(1) The source files are in a single repository (25 points)–**fullfilled** (2) The text of the code is written in the same style (45 points)–**not fullfilled** (3) The text of the code meets the current requirements (30 points)–**fullfilled**
7	Summing up Techology scores	$\mathbf{m}_{Technologies} = \frac{25+30}{100} \cdot 20 = 11\,\textbf{points}$

Table 5 Distribution of points

The result of the assessment in points	PPMMM (Prado Project Management Maturity Model) level
0–20	1 (Initial)
21–40	2 (Known)
41–60	3 (Standardized)
61–80	4 (Managed)
81–100	5 (Optimized)

be carried out on the basis of data on the fullness of weekly or other sprints implemented within the framework of the chosen project methodology, based on data from structured and visualized business processes at the level of management accounting, with separate attention to bottlenecks in processes and optimization of "inputs" and "outputs" in each block. Such an assessment can significantly affect the optimization of the company's activities in terms of minimizing costs and identifying existing or potential risks.

4 Results

To test the proposed methodology, an assessment of the product maturity of the X team engaged in software development (creation of a messenger and chatbots) was carried out. Down below you may take a look at the selected Blocks and the criteria by which the assessment was made.

Development Block:

1. Test cases have been prepared;
2. Automatic testing is used;
3. Checklists have been prepared;
4. Regression testing is used;
5. Integration testing is used (how well the feature interacts in a bundle);
6. Manual testing is used (imitation of user actions);
7. Different devices are used (checking how the program works on different devices);
8. CI CD has been applied (delivery of new modules of the project under development to interested parties (developers, analysts, quality engineers, end users, etc.);
9. Feedback collected;
10. Feedback is regularly updated.

This block provides access to the team's development process and to the maturity level of its development methods. The development block includes criteria such as the quality of the code, the quality of testing methods and the quality of the implementation process. The quality of the code determines the reliability, availability and scalability of the code. Testing shows the level of coverage of test cases and the availability/use of automation tools. This block reflects the quality and maturity level of the most important stage of the creation of the product.

Metrics Block:

1. The results of the analysis of analog products are applied;
2. The work uses a portrait of the target audience;
3. Overview for environment analysis;
4. The value of the created product for the Company's employees is determined and applied;
5. Calculated DAU WAU MAU (Daily/Weekly/Monthly Average Users);
6. Calculated ARPU (Average Revenue Per User);
7. Determined Dev Time;
8. Release Time Is Defined;
9. Cycle Time is Defined;
10. Lead Time Is Defined;
11. Time to Market is defined;
12. Defined Time to Learn;
13. No outstanding tasks at the end of the sprint.

This block allows the team to measure and track metrics collected in the course of work. It is very important in the context of digital transformation, as it gives an idea of the performance of the product and the ability of the team to achieve business goals set. The metrics block includes factors such as DAU, WAU, MAU, ARPU, as well as the time of product release and some others. The number of unaccomplished tasks is monitored separately to improve their setting in the next sprints. The rest of the data arrays are analyzed in the same way to improve the quality of decision making.

Management Block:

1. The described project methodology is used;
2. Project management tools are used (Jira,Trello etc.);
3. A backlog is being maintained;
4. The backlog is being audited (1 time per month);
5. Planning meetings are held;
6. Metrics are used in project management;
7. The team has all the necessary specialists;
8. The team conducts document flow in electronic form;
9. There is a single guideline for solving standard tasks;
10. Feedback from team members is collected;
11. A motivation system is used;
12. Tasks are prioritized;
13. Team members use an electronic signature.

The block includes criteria to evaluate the team management such as a well-defined team structure, the presence or absence of a project management methodology with described and transparent tools. It is important to evaluate the management block, as it can have a significant impact on the team's ability to release the final product on time and within the budget.

Interaction Block:

1. The process of interaction with vendors has been established (technical specification, deadlines, priorities, setup meetings, established communication);
2. Team building is carried out [23];
3. There are no serious conflict situations;
4. There is a system for transferring experience within the team;
5. There is a team chat in the messenger;
6. Brainstorming is underway;
7. Corporate culture is defined.

This Block reflects the ability of cross-functional teams to interact with other teams and stakeholders. It includes factors such as interactions with vendors, external contractors, teamwork and conflict resolution inside it.

It is important to evaluate this block, as it can help the team build better relationships with stakeholders and other teams, which can have a positive impact on the final product.

As a result of interviews with team members and expert evaluation, the results given in Table 6 were obtained.

The scores obtained during the evaluation were summed up and the resulting Total value corresponds to Level 4 (Managed) in accordance with PPMMM (Prado Project Management Maturity Model) level (Table 5).

To illustrate the results obtained and identify weaknesses that require attention, a radar chart (Fig. 1), reflecting the numerical values of the results of the assessment was created.

Table 6 Evaluation results

Block	The percentage of the maximum block weight gained, %
Development	74,5
Metrics	87,5
Management	90
Interaction	67,5
Total	**79,875**

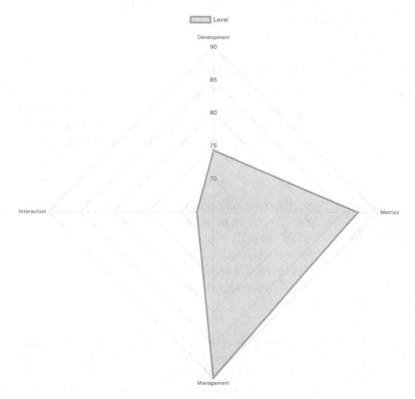

Fig. 1 Distribution of the evaluation results on the radar chart

The diagram clearly shows the possible areas of growth and development of the team. Let's take a closer look at each of the evaluated blocks.

To begin with, it is worth noting the high level of project management in the team. The team selected for evaluation works in a large IT company and follows clear and transparent guidelines adopted by the company, in this regard, the high result of the team reflects preliminary expectations and serves as a good indicator, since the management unit is the link between the others.

It can be noted that the Metrics block is also at a high level, which corresponds to a high level of management, since the criteria evaluated in it are clearly fixed by the project manager and their value is clear to all team members.

The development unit is at a good average level and meets most of the current requests from the customer and the market as a whole, but not all possible tools are used by the development team. The reason for this may be that there is no need to implement additional tools, since the team is not large.

The interaction block turned out to be the least mature. To determine the cause of the situation, the company's management should pay attention to the formation and strengthening of corporate culture, and it is also possible to adapt approaches to project management taking into account the soft skills of cross-functional teams.

In general, it is worth noting that the result of 79.875% indicates a high level of professionalism. According to PPMMM, the team is at the penultimate maturity level (Table 5) and is able to effectively manage its processes. Nevertheless, the announced challenges are guidelines for the further development of the company. With the help of assessment tools, the company will be able to better understand itself, but in order to remain at the current level of maturity, and even more so to become more mature, it needs to constantly develop each of the Blocks, gaining criteria that correspond to global trends and its own development strategy.

In addition to identifying the strengths and weaknesses of the team when assessing its maturity, the tools offered for implementation can serve as additional tools for HR specialists. The summary results of the evaluation by blocks can be used when building the development track of an individual employee. Based on the results of the evaluation, you can see which of the criteria within the blocks are insufficiently developed and decide on additional professional training, on the introduction of additional incentive mechanisms for employees or on optimizing the staffing of project teams or teams that ensure the operational activities of the enterprise.

It is also important to note here that such an approach at the current stage of technology development is only an additional tool for decision-making and does not replace all other existing ones.

5   Conclusion

As a result of testing the methodology on specific product teams, it can be concluded that the proposed method is quite flexible and is able to accurately reflect weaknesses that require improvement within the team or at the level of the company's processes as a whole. Understanding the specific weaknesses of the team is an important step and a need for digital transformation. The data obtained as a result of the maturity assessment can be predictors for training a digital twin model. Such data allows us to decompose existing levels of business processes and allow us to build more detailed strategic models in the context of digitalization and digital transformation. The proposed tools will improve the quality of decisions made within the company and increase operational efficiency.

The issue of the global trend towards digital transformation remains important and relevant. As already noted in the chapter, digital transformation requires companies to fully restructure existing business processes and build a system in which the process of collecting feedback from the market and an individual client is optimized and automated, as well as analyzing data and creating new options for building business processes and positioning strategies based on this data. Evaluation of the product development process can help companies independently understand their strengths and weaknesses, as well as decompose processes into their component parts and understand cause-and-effect relationships.

However, the development of a comprehensive methodology for such an assessment may be subject to adaptation and, as a result, include new indicators for evaluation. The problem of a high human factor at the expert level of decision-making remains relevant both when choosing a pool of values and when determining the weight of each of the blocks. If such criteria are determined at the level of the product owner or project manager, it may be difficult to objectively compare the results obtained. Therefore, it is recommended to strive to unify the selected evaluation criteria by blocks, in accordance with the priorities and requirements of the company's top management or top management. However, the model proposed for implementation has no restrictions on the frequency of criteria updates and the number of criteria or blocks, provided that the remaining evaluation rules are followed.

References

1. Pulse of the Profession 2017|PMI. https://www.pmi.org. Accessed 18 May 2023
2. Kerzner H. (2019) Using the project management maturity model: strategic planning for project management. H. Kerzner, Wiley, pp 313
3. Tessaro JA, Harms R, Schiele H (2023) How startups become attractive to suppliers and achieve preferred customer status: Factors influencing the positioning of young firms. Ind Mark Manag 113:100–115
4. Berg H et al (2023) Successful IT projects–A multiple case study of benefits management practices. Procedia Comput Sci 219:1847–1859
5. Lermen FH et al (2023) Does maturity level influence the use of Agile UX methods by digital startups? Evaluating design thinking, lean startup, and lean user experience. Inf Softw Technol (154):107
6. Martinsuo M, Ahola T (2022) Multi-project management in inter-organizational contexts. Int J Project Manag 7(40):813–826
7. Domingues L, Ribeiro P (2023) Project management maturity models: proposal of a framework for models comparison. Proc Comput Sci 219:2011–2018

8. Elhusseiny HM, Crispim JA (2023) Review of Industry 4.0 maturity models: adoption of SMEs in the manufacturing and logistics sectors. Proc Comput Sci (219):236–243
9. Kieroth A et al (2022) Investigation on the acceptance of an Industry 4.0 maturity model and improvement possibilities. Proc Comput Sci (200):428–437
10. Miloslavskaya N, Tolstaya S (2022) Information security management maturity models. Proc Comput Sci 213:49–57
11. Simetinger F, Basl J (2022) A pilot study: an assessment of manufacturing SMEs using a new Industry 4.0 maturity model for manufacturing small- and middle-sized enterprises (I4MMSME). Proc Comput Sci (200):1068–1077
12. Silva PP et al (2021) Maturity model for collaborative R&D university-industry sustainable partnerships. Proc Comput Sci (181):811–817
13. Kim S et al (2022) Organizational process maturity model for IoT data quality management. J Ind Inf Integr (26):100–256
14. Komarov AV, Komarov KA, Shurtakov KV (2021) Using the methodology for the comprehensive assessment of scientific and technological projects to estimate risks of their failure. Econ Sci 1(7):19–38
15. Silvius G, Marnewick C (2022) Interlinking sustainability in organizational strategy, project portfolio management and project management a conceptual framework. Proc Comput Sci 196:938–947
16. Komarov AV et al (2020) Specialized tools to evaluate the potential of a R&D project team. Econ Sci 1–2(6):75–87
17. Dolat D et al (2017) An assessment for IT project maturity levels. Int J Inf Technol Project Manag 2(8):1–16
18. Derenskaya Y (2017) Organizational project management maturity. Baltic J Econ Stud 2(3):25–32
19. Guo K, Zhang L (2022) Multi-objective optimization for improved project management: current status and future directions. Autom Constr 139:104–256
20. Silakova LV, Nikishina A (2021) Digital transformation of telecom providers management customer system: a process research and effects assessment. In: ACM international conference proceeding series, pp 81–90
21. Clemente M, Domingues L (2023) Analysis of project management tools to support knowledge management. Proc Comput Sci 219:1769–1776
22. Hein-Pensel F et al (2023) Maturity assessment for Industry 5.0: a review of existing maturity models. J Manuf Syst (66):200–210
23. Daniel PP, Daniel PA, Smyth H (2022) The role of mindfulness in the management of projects: potential opportunities in research and practice. Int J Project Manag 7(40):849–864

Support for Management Decision-Making Based on the "HAM" Method and the DL "Random Forest" Model to Increase Company Efficiency

Nikolay Lomakin⬤, Maxim Maramygin⬤, Alexey Polozhentsev⬤,
Julia Polozhentseva⬤, Pavel Kravchenya⬤, and Galiya Rakhmankulova⬤

Abstract The chapter discusses the theoretical foundations support for management decision-making based on the HAM–method and the DL "Random forest" Model to increase the company's efficiency. The relevance of the study lies in the fact that in modern conditions it can be difficult for a customer enterprise to choose a reliable contractor for construction work, which leads to an increase in the cost of construction and the appearance of lost profits for production. Practice shows that such losses are significant and have a significant impact on production efficiency. Since many factors affect the company's performance under conditions of market uncertainty, it seems appropriate to use the Random Forest deep learning model to predict the net profit of an enterprise. The novelty of the study lies in the proposed approach, which involves the use of the mathematical apparatus of the artificial intelligence system Model DL "Random Forest" to predict the company's net profit in the process of supporting management decision-making related to the selection of reliable contractors by the HAM method and evaluating the effectiveness of this choice. A hypothesis has been put forward and proved that using the AI system it is possible to provide support for making a decision on the choice of a contractor and obtain a forecast of the company's net profit.

N. Lomakin (✉) · P. Kravchenya
Volgograd State Technical University, ave. V.I. Lenina, 28, Volgograd 400005, Russia
e-mail: tel9033176642@yahoo.com

M. Maramygin
Ural State University of Economics, 8 Marta St., 62, Yekaterinburg 620144, Russia

A. Polozhentsev
Voronezh State University, University Square, 1, Voronezh 394018, Russia

J. Polozhentseva
Southwestern State University, st. 50 Years of October, 94, Kursk 305040, Russia

G. Rakhmankulova
Volzhsky Polytechnic Institute (branch) of Volgograd State University, Engels str., 42a, Volzhsky 404111, Russia

© The Author(s), under exclusive license to Springer Nature Switzerland AG 2023   85
A. Bencsik and A. Kulachinskaya (eds.), *Digital Transformation: What is the Company of Today?*, Lecture Notes in Networks and Systems 805,
https://doi.org/10.1007/978-3-031-46594-9_6

Keywords HAM–method · Model DL "Random forest" · Reliable contractor ·
Efficiency · AI system · Hybrid intelligent systems

1 Introduction

In the presented work, such research methods were used as: monographic, analytical,
statistical, the method of analysis of hierarchies "HAM", as well as the artificial
intelligence system Deep learning model "Random Forest". Based on statistical data
and the results of the work of Caustic company for the period from 2010 to 2022
years, a selection of the main macroeconomic indicators reflecting the dynamics of
the development of the economy and the company in question has been formed.

The purpose of the work is to use the HAM method to select reliable contractors for construction and installation work at the enterprise, and thereby reduce the
annual cost of building construction, and ultimately calculate the forecast value of
the company's net profit. To achieve the goal, the following tasks were set and solved:
(1) to explore the theoretical foundations for supporting managerial decision-making
in modern conditions; (2) to create a Digital Bot script that would allow to collect the
necessary data from the Websites of potential contractors, provide an opportunity for
experts to remotely give their assessments and form a list of five reliable contractors;
(3) calculate the forecast value of net profit based on the results of the company's
work for the year.

Caustik is one of the leading enterprises of the chemical industry in Volgograd
[1]. The relevance of the study lies in the fact that the problems that arise during
the interaction "enterprise–contractor", based on the use of traditional methods in
classical formats (joint meetings, teleconferences, the use of CRM and ERP systems)
can be solved using modern digital services in the format "Digital Ecosystem".
In the face of increasing complexity in the functioning of the construction sector
as a system: customer → contractor → subcontractor → auditors → commission
experts, it becomes difficult to control the timeliness of execution of orders, manage
a huge number of respondents, an increasing flow of information, one has to deal
with conflicts of office work, approvals in various kinds authorities and so on. It is
important to study the trends in the development of relationships between participants
in the investment process using digital technologies in modern conditions.

An analysis of the work of the Caustic enterprise showed that the relationship
between the customer and contractors can be built using various schemes. So, for
example, the distribution of functions between the participants is determined by
the contractual terms and the schemes of interaction between the participants can
be varied. Since contractual relations are built on the basis of interaction between
the customer company and numerous construction participants, this requires the
automation of business processes and support for managerial decision-making to
ensure the efficient operation of the company. Despite the processes of improving the
relationship between the customer company and construction contractors, problems

remain. The analysis showed that among the main problems during construction, the following can be noted:

- rise in the cost of construction objects (for various reasons, for example, an improperly conducted survey of a land plot for construction);
- postponing the completion of the commissioning of the facility in the direction of increasing the construction time;
- the emergence of controversial issues in the relationship "Customer-Contractor", often due to improper fulfillment of obligations under the contract.

Various material losses of the Customer due to an increase in the cost of construction and installation works are presented below (Table 1).

The following entities may be involved in the process of interaction between participants in contractual relations regarding the implementation of investment projects in the construction of facilities of a private company-customer: designers, general contractors, contractors, subcontractors, equipment suppliers, a bank, representatives of the land owner, specialists for examination, expert analysts and others.

The relationship between the participants is constantly in dynamics. The interaction of participants in contractual relations is ensured by the management of construction projects. Investment (construction) project management is a way of organizing production, aimed at the timely achievement of a one-time, non-recurring goal with the optimal use of available resources. The Hierarchy Analysis method was not used in the evaluation of contractors at the enterprise.

In our case, we will consider the AHP using the example of a two-level system. Let there be a set of N solutions (objects) An ($n = 1, 2, ... N$) from which to choose the best solution or the best object. To evaluate each of the objects, the same set B_m of M comparison criteria is used $m = 1, 2, ... M$. To build the final assessment of solutions, you must first compare the criteria, and then pairwise the solutions to each other for each of the criteria.

Table 1 Rise in the cost of construction objects Caustic

Year	Description	Cost of appreciation, thousand rubles	Cause
2020	Construction of Object № 4	12.5	Hidden foundations were revealed due to under-examination of the territory
2021	Construction of Object № 4, 5	18.8	Change in the estimate due to force majeure circumstances (raise in the cost of materials due to sanctions)
2022	Construction of Object №5	7.4	Disadvantages of marking by a third-party organization, some factors remained unaccounted for
	Total	38.7	

We have to compare criteria that are usually not of equal importance. A pairwise comparison of the criteria is carried out on a qualitative scale, followed by conversion to scores:

(1)   the same, indifferent $= 1$;
(2)   slightly better (worse) $= 3$ (1/3);
(3)   better (worse) $= 5$ (1/5);
(4)   much better (worse) $= 7$ (1/7);
(5)   fundamentally better (worse) $= 9$ (1/9).

With an intermediate opinion, they use With an intermediate opinion, intermediate scores of 2, 4, 6, 8 are used. The results are entered into a table. In the same way, they create matrices of pairwise comparison of objects (solutions) by criteria [2].

To fulfill the consistency conditions in the matrices of pairwise comparisons, reciprocal values are used. If a_{ij} have an attitude criterion i to criterion j, then $a_{ji} = 1/a_{ij}$.

If there are quantitative criteria for assessing any quality ϖ_j, then the comparison result for two objects is found simply by dividing $\varpi_j / \varpi_k = a_{jk}$. Accordingly, the consistency condition should be automatically satisfied.

$$\frac{\omega_j}{\omega_k} \frac{\omega_k}{\omega_m} = a_{jk} * a_{km} = a_{jk}/a_{mk}, \tag{1}$$

If pairwise comparisons are made using a scale of qualitative preferences, then condition (1) may be violated. In this case, the estimates will be violated by automatic consistency and you need to have a mathematical criterion that confirms that the consistency is maintained at an acceptable level. To check, we can verify that, with complete consistency and condition (1), the following equality holds:

$$A_\omega = n\omega, \tag{2}$$

where $A = (a_{jk})$, $\omega = (\omega_1, \omega_2, \ldots \omega_n)$, n–number of paired comparisons.

If condition (1) is violated, then the maximum eigenvalue of Eq. (2) will be different from n and the deviation value can be taken as a measure of consistency in the form of a coefficient

$$C = \frac{\lambda_{max} - n}{n - 1}, \tag{3}$$

where λ_{max}–the maximum eigenvalue of the matrix A. Consistency matrix elements is taken at $C < 0.1 \div 0.2$.

If all comparison matrices both for comparing criteria and for comparison of objects according to the criteria are completed and agreed, you can proceed to calculations. It is necessary to make integral estimates of M criteria by the matrix.

$$x_i = \sum_{l=1}^{M} a_{il}, \; S_A = \sum_{i=1}^{M} x_i, \; q_i = \frac{x_i}{S_A}, \tag{4}$$

Normalizing the result by dividing by the sum S_A provides distribution of marks in shares 1. The same procedure must be done with comparison matrices N of objects for each of the criteria. As a result we get M columns (according to the number of criteria) of N elements (according to the number of objects comparisons) that can be combined into a matrix y_{ni}. After that you can calculate the normalized estimates of each of the N objects (solutions) taking into account M criteria weights x_i calculated earlier by formula (4):

$$c_n = \sum_{l=1}^{M} x_i y_{ni}, \; S = \sum_{n=1}^{N} c_n, \; p_n = \frac{c_n}{S}, \tag{5}$$

Comparison of estimates obtained in this way p_n allows for the greatest of them to choose the best object (solution). To automate the calculations, the calculations were carried out in a program in the Python language. The study of the features of the system of relationships between the participants in the investment process in construction based on innovations in the context of digital transformations of business processes in the construction sector, as well as current trends and development prospects, is important.

2 Literature Review

Many scientific studies are devoted to the study of the problems associated with the formation of relationships between the participants in the construction process. So, for example, Kvartina V.V. considered trends in the formation of a system of relationships between participants in the investment process in construction [3]. Kovalchuk S.I. revealed an algorithm for the formation of a system of relationships between participants in the investment process [4]. The solution of most problems, as the analysis showed, is associated with the need to organize the search, evaluation and selection of a reliable contractor with the involvement of external independent experts, which would minimize the risk of an increase in the cost of building an object. Thus, it is required to develop a new approach in the evaluation of the contractor by the method of analysis of hierarchies (hereinafter AHP).

At the moment, the selection procedure is based on the methodology provided for in the conduct of tenders by the customer organization, the criterion of which is the minimum price for the construction contract. The AHP method ensures the objectivity of the choice of a contractor, which eliminates the possibility of a corruption component.

The organization of the work of experts can be built on a remote basis using the Robo-bot proposed by the authors of the chapter. Since the rise in construction costs is not the only factor affecting the amount of net profit and it is necessary to take into account many other factorial features, it seems appropriate to use the Random Forest Deep Learning Model to predict this parameter, which reflects the efficiency of the company as a whole. Research shows that artificial intelligence technologies are increasingly being used. So, for example, Bataev A.V., Gorovoy A.A. and Denis Z. conducted a comparative analysis of the use of neural network technologies in the world and Russia, stating that the prerequisites for the rapid use of artificial intelligence are the processes of large-scale development of information and communication technologies, an increase in the volume of processed information, as well as the development of production capacities of computers in data processing centers, and other factors [5].

Assessment and reduction of financial risk remain in the focus of attention of entrepreneurs and enterprises in the real sector of the economy. So the team of authors Shokhnekh A., Lomakin N., Glushchenko A., Sazonov S., Kovalenko O. and Kosobokova E. proposed a digital neural network for managing financial risks in business through real options in the financial and economic system [6]. Support for managerial decision-making is always associated with risk, the risk of financial losses, its assessment and minimization. In the deep risk model proposed by Hengxu Lin, Dong Zhou, Weiqing Liu and Jiang Bian, a deep learning solution is proposed to analyze latent risk factors while improving the estimation of the covariance matrix. Experiments were carried out on stock market data and demonstrated the effectiveness of the proposed solution [7].

The assessment of financial risks using the VaR model provides high performance to support managerial decision-making in the real sector of the economy in the financial sector. A group of scientists consisting of Kei Nakagawa, Shuhei Noma and Masaya Abe proposed an approach based on the use of the RM-CVaR model. It is known that dispersion is the most fundamental measure of risk that investors seek to minimize, but it has a number of disadvantages. Notional Value at Risk (CVaR) is a relatively new risk measure that overcomes some of the shortcomings of well-known variance risk measures and has gained popularity due to its computational efficiency. CVaR is calculated as the expected value of a loss that occurs beyond a certain probability level (β) [8]. The use of artificial intelligence is increasingly evident in the use of robotic advisors in business. Katherine D'Hondt, Rudy De Winn, Eric Gizels, and Steve Raymond conducted a study on the use of an AI-enabled "alter ego" system in robotic investments. The authors introduced the concept of "AI Alter Ego", which are shadow robot investors [9]. One of the promising areas is the use of deep neural networks in business. For example, Krzysztof R. et al. proposed neural risk assessment in networks with unreliable resources [10]. According to the authors, it is advisable to use an algorithm based on GNN, which is trained only on random graphs generated using the Barabashi-Albert model. Clarkson J. et al. proposed the DAMNETS neural network, which is a deep generative model for Markov network time series. Time series networks are found in many fields such as trade networks in economics [11].

The use of generative models is effective for Monte Carlo estimation and improvement of the original dataset used, which is of interest for model fitting. Bingyang H.

studied the estimation of the expected values of a function from time data. He approximated test functions with neural networks in order to prove sample complexity using Rademacher complexity [12]. Neufeld A. suggested using robust data-driven statistical ingestion strategies using deep neural networks [13]. The conducted research makes it possible to obtain an increment of knowledge that allows to close the scientific gap regarding the support of managerial decision-making in modern conditions. At the same time, the scientific approach is based on two approaches, the use of the Hierarchy Analysis Model (HAI), as well as the artificial intelligence system–the Random Forest Deep Learning Model.

3   Materials and Methods

This study was carried out using such methods as: monographic, analytical, statistical, the method of analysis of hierarchies "MAH", as well as the artificial intelligence system Deep learning model "Random Forest", based on statistical data and the results of Caustic's work for the period from 2010 to 2022 years. A selection of the main macroeconomic indicators was formed, reflecting the dynamics of the development of the Russian economy, as well as the performance of the company in question.

3.1   The Hierarchy Analysis Method (HAM) for Selecting a Reliable Building Contractor

Modern enterprises are complex economic systems. To analyze economic systems, T. Saaty developed the Hierarchy Analysis Method (HAM), which is a systematic procedure for evaluating a system using a hierarchical representation of the elements that define the essence of the problem.

In the course of the study, a model was formed for evaluation by the HAM method and a comparison was made of a system of five contractor companies according to the criteria: size, financial stability, own resources, reliability, reputation (Fig. 1).

The method consists in decomposing the problem into simpler components and processing the decision maker's sequence of judgments based on paired comparisons. The relative importance of elements in a hierarchy is expressed numerically. The resulting values are estimates in the ratio scale.

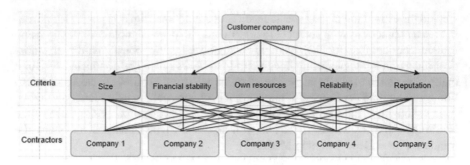

Fig. 1 Formation of the HAM model

3.2 Digital-Bot for Automatic Collection of Information in Order to Select Reliable Construction Contractors

Since in modern conditions almost any information about the company is posted on its Web site, it seems appropriate to develop the Digital-Bot program to collect information about construction contractors, their ranking and based on the hierarchy analysis method (HAM). Below is a diagram of contractor companies by their rating, which is based on the "Revenue" parameter (Fig. 2).

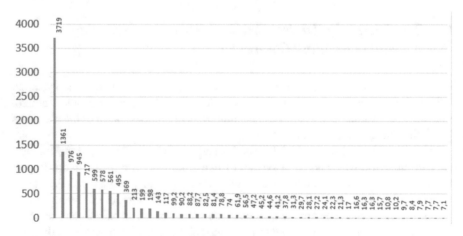

Fig. 2 Diagram of construction companies in the Volgograd region by revenue (million rubles)

3.3　The Random Forest Deep Learning Model for Predict the Net Profit

The initial data for the formation of the Random Forest deep learning model is presented below (Table 2).

To predict the net profit of the customer company, the Random Forest deep learning model was formed. In order to form a dataset of the model, external and internal factors were selected that affect the final result of the company's work–net profit. The following factors were considered: Year; GDP, billion rubles (GDP); Dollar exchange rate, rub. (USD); Investments in fixed assets in GDP, % (Investments); Rise in the cost of services of building contractors, thousand rubles (Increasing costs); Revenue of Caustic, thousand rubles (Revenue); Assets of Caustic, thousand rubles (assets); Intangible assets of Caustic, thousand rubles (Intangible assets); Forecast of net profit of company Caustic, billion rubles (Net profit forecast). To obtain the predictive value of the effective feature–the value of the company's net profit, using the artificial intelligence system, we will form a data set of the model.

4　Results

As a result, of the study, forecast values of the volume of overdue loans received in two different ways were obtained: based on a correlation-regression model. a neural network. Using financial macroeconomic indicators, which were formed with Central Bank data, made it possible to build a correlation-regression model, as well as a perceptron artificial intelligence system.

4.1　The Hierarchy Analysis Method (HAM) for Selecting a Reliable Building Contractor

The process of forming the (HAM) method begins with pairwise comparisons of criteria (Fig. 3).

The calculation of the significance coefficients by the hierarchy analysis method makes it possible to calculate a number of parameters as comparison criteria. Summary parameters for the alternatives are given below (Table 3).

Thus, the positions were distributed as follows: (1) Limited Liability Company (LLC) Construction Company South–0.311; (2) Absolut Group LLC–0.303; (3) LLC "Volga-BUILD engineering"–0.232; (4) LLC "LLC "ATN""–0.091; (5) Volgtranstroy LLC–0.063. Based on the results obtained by the MAI, Caustic should change the list of contractors, terminate contracts with contractors with a low rating of experts and conclude contracts with contractors with a high rating. This will reduce

Table 2 Initial data for the formation of the Random Forest deep learning model

Year	GDP, billion rubles	Dollar exchange rate, rub	Investments in fixed assets in GDP, %	Rise in the cost of services of building contractors, thousand rubles	Revenue of Caustic, thousand rubles	Assets of Caustic, thousand rubles	Intangible assets of Caustic, thousand rubles	Net profit of company Caustic, billion rubles
2022	153,435	70.3	20.2	7.4	32,459,133	27,712,102	11,365	6,773,452
2021	131,015	73.7	21.2	18.8	29,469,846	23,687,263	57,341	5,088,906
2020	107,315	73.8	16.5	12.5	21,805,092	8,591,084	24,836	1,146,733
2019	109,242	62.0	20.6	25.2	20,857,785	20,703,477	17,585	1,894,435
2018	103,862	69.8	20.6	60.0	20,677,612	19,356,329	16,561	3,099,970
2017	91,843	57.6	21.4	33.3	17,834,688	17,322,311	15,065	2,082,978
2016	85,616	61.3	21.2	15.4	17,834,688	15,280,139	14,461	2,299,848
2015	83,087	73.6	20.0	56.9	12,548,510	14,363,951	15,262	1,631,726
2014	79,030	55.9	20.5	154.1	12,548,510	15,395,972	16,630	320,695
2013	72,986	32.9	21.2	61.0	11,901,775	16,057,304	14,563	510,967
2012	68,103	30.6	20.9	53.9	9,829,008	15,568,530	14,873	477,796
2011	60,114	32.2	20.7	81.4	7,756,241	14,608,550	16,038	444,625
2010	44,491	30.2	20.6	30.8	6,869,435	9,687,733	16,807	629,711

```
1   # Contractors
2   alt1 = ['Company 1','Company 2', 'Company 3','Company 4','Company 5']
3   # Criteria: Size
4   EX = np.mat([[1.0, 9.0,  8.0, 8.0, 1.0],
5   [1./9., 1.0, 9.0, 8.0, 9.0],
6   [1./8, 1./9., 1.0, 8.0, 5.0],
7   [1./8., 1./8., 1./8.,1.0, 9.0],
8   [1.0, 1./9., 1./5.,1./9., 1.0]])
9   #print ('Опыт:')
10  print(EX)
```

```
[[1.         9.        8.        8.        1.       ]
 [0.11111111 1.        9.        8.        9.       ]
 [0.125      0.11111111 1.       8.        5.       ]
 [0.125      0.125      0.125    1.        9.       ]
 [1.         0.11111111 0.2      0.11111111 1.      ]]
```

Fig. 3 Script of the program for the formation of the HAM model (fragment)

Table 3 Calculation of significance coefficients by the HAM method

Alternative	Size	Financial stability	Own resources	Reliability	Reputation	Total
Company 1	0.234	0.286	0.334	0.501	0.586	0.303
Company 2	0.234	0.571	0.156	0.159	0.255	0.311
Company 3	0.266	0.143	0.510	0.038	0.069	0.232
Company 4	0.117	–	–	0.075	0.038	0.063
Company 5	0.148	–	–	0.226	0.052	0.091
NW	12.172	5.253	4.573	6.246	6.138	–
CI-consistency index	1.793	1.126	0.787	0.311	0.285	–
RC–relative consistency	1.601	1.942	0.356	0.277	0.254	–

the volume of annual price increases. The expected economic effect will be 15,180 thousand rubles.

4.1.1 Digital-Bot for Automatic Collection of Information for the Hierarchy Analysis Method (HAM)

The program script is hosted in the "cloud" Collab by Google, written using the Python language. The block diagram of the bot is shown in Fig. 4.

Digital-Bot for collecting information about construction contractors, their ranking and selection based on the hierarchical analysis method (HAM), was successfully formed (Fig. 5).

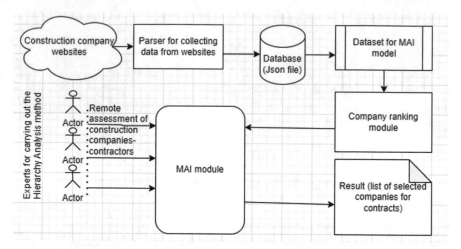

Fig. 4 Digital-Bot for collecting information about construction contractors, their ranking and selection based on the Hierarchy Analysis Method (HAM)

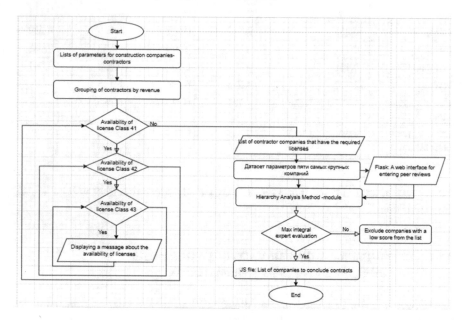

Fig. 5 Digital-Bot algorithm for choosing the optimal contractor

The following modules are part of the Digital-Bot Flowchart for collecting information about construction contractors.

(1) A module that contains http:// hyperlinks to the websites of companies that provide a list of required services provided properly.
(2) Parser, a program that collects the necessary information from the websites of the right companies.
(3) Database (Jason-file), store the collected information systematized in an appropriate way.
(4) Dataset–a file (with the.txt extension) that contains data that is then used to rank companies for inclusion in the HAM model.
(5) The HAM module, in which the necessary calculations are carried out in automatic mode, accepting estimates (parameters from 0 to 10) of experts who remotely assign values to the corresponding parameters.
(6) Result–represents a file in the form of a list of companies that have been assessed by experts, containing integral values–coefficients calculated on the basis of the values of the initial values and assessments of experts according to the assessment criteria used based on the hierarchy analysis method (HAM).

Parser–a program or service for searching data according to certain rules. Processes information according to specified criteria and displays it in a structured form. The input conditions can be a key phrase or any other sequence of characters, as well as object characteristics–file type and size, geolocation, etc. As a result of the implementation of the proposed measures in the company Caustic, the expected increase in income may amount to 28,880 thousand rubles. Thus, the estimated increase in profit from the system of improvements can be 15,180 thousand rubles, and the cost-effectiveness ratio for the implementation of improvements will be 1.108.

4.2 Random Forest Deep Learning Model

To obtain the predictive value of the effective feature–the value of the company's net profit, using the artificial intelligence system, we will form a data set of the model. The neural network data set was successfully formed (Table 4).

The settings of the hyperparameters of the model included, for example, the number of estimators 5, the maximum depth of the tree 10. As a result, of using the GridSearchCV library, the parameters of the best tree were calculated (Fig. 6).

The Best Ensemble Decision Tree Random Forest Deep Learning Models is shown below (Fig. 7).

Table 4 Neural network dataset

	GDP	USD	Investments	Increasing costs	Revenue	Assets	Instangible assets	Net profit forecast
0	153,435	70.3	20.2	7.4	32,459,133	27,712,102	11,365	6,773,452
1	131,015	73.7	21.2	18.8	29,469,846	23,687,263	57,341	5,088,906
2	107,315	73.8	16.5	12.5	21,805,092	8,591,084	24,836	1,146,733
3	109,242	62.0	20.6	25.2	20,857,785	20,703,477	17,585	1,894,435
4	103,862	69.8	20.6	60.0	20,677,612	19,356,329	16,561	3,099,970

```
▸          GridSearchCV
▸ estimator: RandomForestRegressor
   ▸ RandomForestRegressor
```

```
[33]  1  gs.best_score_
-328779.69999999995
```

```
[34]  1  gs.best_params_
{'criterion': 'poisson', 'max_depth': 10, 'n_estimators': 5}
```

Fig. 6 The parameters of the best tree

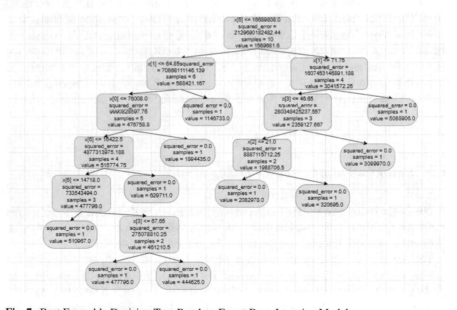

Fig. 7 Best Ensemble Decision Tree Random Forest Deep Learning Models

The matrix of paired correlation coefficients between factorial and resultant features is presented below. As studies have shown, the following factors have a strong positive effect on the value of the enterprise's net profit: Revenue 0.92; GDP 0.885; Assets 0.833. A negative impact on net profit is exerted by Increasing costs −0.525 (Table 5).

Actual and predictive values of the resulting feature, which are randomly generated by the DL Model (Table 6).

Substituting the values of factor indicators for 2022 into the model, the forecast parameter of the effective indicator was calculated–the value of the net profit of Caustic is 3749506.95925706 billion rubles, while this parameter was 4480 billion rubles. While the actual value for 2022 was 6,773,452 thousand rubles, that is, the forecast turned out to be -80.6% less. The use of the skikitlern and pandas libraries made it possible to obtain regression coefficients.

The linear regression equation has been obtained.

$$Net\ profit\ forecast = -131.408765 * GDP + 61185.53979$$
$$* USD - 292491.437621 * Investments - 8091.838754$$
$$* Increasing\ costs + 0.316342 * (Revenue + 0.334116$$
$$* Assets + 31.90103 * assets) \qquad (6)$$

Table 5 Matrix of paired correlation coefficients

	GDP	USD	Investments	Increasing costs	Revenue	Assets	Instangible assets	Net profit forecast
GDP	1.000	0.784	−0.190	-0.452	0.983	0.767	0.368	0.885
USD	0.784	1.000	−0.369	−0.303	0.774	0.382	0.345	0.628
Investments	−0.190	−0.369	1.000	0.182	−0.168	0.429	−0.027	0.090
Increasing costs	−0.452	−0.303	0.182	1.000	−0.534	−0.207	−0.220	−0.525
Revenue	0.983	0.774	−0.168	−0.534	1.000	0.371	0.441	0.902
Assets	0.767	0.382	0.429	−0.207	0.731	1.000	0.229	0.833
Instangible assets	0.368	0.345	−0.027	−0.220	0.441	0.229	1.000	0.354
Net profit forecast	0.885	0.628	0.090	−0.525	0.902	0.833	0.354	1.000

Table 6 Actual and predictive values of the effective feature

	Actual	Predicted	Delta_persent
6	2,299,848	320,695.0	86.055818
0	6,773,452	5,088,906.0	24.869830
7	1,631,726	1,146,733.0	29.722699

Using the pandas libraries, numpy will generate a file Sample (Table 7).

Using the pandas, numpy libraries will generate a file. As a result of using the parameters uploaded to the Sample file, a predictive value was formed in the DL RF Model. We get the predictive value of the effective feature:3,749,506.95925706. Whereas the actual value for 2022 was 6,773,452 thousand rubles, or −80.6% less. As a result of substituting the Sample file parameters into the linear regression equation, the forecast was obtained: 3,749,506.95925706. It coincides with the forecast from DL_RF. Results of assessing the quality of the model presented in Table 8.

Model Quality Options are calculated as follows.

MAE is a metric that tells us the average absolute difference between the predicted values and the actual values in the dataset.

$$MAE = \frac{1}{n} * \sum |y_i - \hat{y}_i| \tag{7}$$

where: Σ is the symbol that means "sum";

y_i–observed value for the *i-th* observation;

$\hat{y}_i$–predicted value for the *i-th* observation;

n -is the sample size.

The lower the MAE, the better the model fits the dataset.

Mean Squared Error is used when it is necessary to emphasize large errors and choose a model that gives less large errors. Large error values become more noticeable due to the quadratic dependence. MSE is calculated using the formula:

$$MSE = \frac{1}{n} * \sum_{1}^{n} (y_{i-}\hat{y}_i)^2, \tag{8}$$

Table 7 File Sample for get the predictive value

	GDP	USD	Investments	Increasing costs	Revenue	Assets	Instangible assets
0	153,435	70.34	20.2	7.4	32,459,133	27,712,102	11,365

Table 8 Results of assessing the quality of the model

Model Quality Options	The share of the test sample in the dataset, 0.20	The share of the test sample in the dataset, 0,30
Mean Absolute Error (MAE)	1,131,354.33	714,688.25
Mean Squared Error (MSE)	1,890,804,710,507.6667	817,177,030,670.75
Root Mean Squared Error (R2)	1,375,065.3477226698	903,978.4459104929

where: n–the number of observations on which the model is built and the number of forecasts,

 y_i–the actual value of the dependent variable for the i-th observation,

 $\hat{y}_i$ is the value of the dependent variable predicted by the model.

Standard deviation (RMSD) or standard error (RMSE) otherwise R2 is a commonly used measure of the difference between values (sample or population values) predicted by a model or estimator and observed values. RMSD is the square root of the second sampling point of differences between predicted values and observed values. Root Mean Squared Error is calculated simply as the square root of the MSE:

$$RMSE = \sqrt{\frac{1}{n} * \sum_{1}^{n}(y_i - \hat{y}_i)^2}, \qquad (9)$$

Data analysis allows us to conclude the average level of forecast accuracy. So the level of the average absolute error of the model was 714,688.25 thousand rubles. in the case where the share of the test sample in the dataset is 0.30. At the same time, the MAE error increased to 1,131,354.33 thousand rubles, or 1.58 times, if the ratio of the test sample is reduced to 0.20.

5 Discussion

Correlating the results obtained with the questions posed in the introduction, we can say that for further research it is desirable to use models of neural networks of other classes, for example, CNN. A convolutional neural network is a deep learning algorithm that can take input parameters, assign importance (digestible weights and biases) to different areas, objects, depending on the purpose of the study. Or, replacing the resulting features with others with higher regression coefficients, select the hyperparameters of the random forest trees in such a way as to minimize the mean absolute error (MAE).

It should be noted that the results obtained in this chapter are based on the experience of past research by the authors. For example, Lomakin et al. offered an artificial intelligence system has been proposed for processing big data to determine the value of innovative products in a digital economy [14]. This study provides an increment of scientific knowledge that allows closing the scientific gap in terms of identifying and assessing the influence of factors that ensure support for management decision-making in order to increase the net profit of the investigated company Caustic in modern conditions. The development of the computing power of modern cloud clusters has made it possible to use modern neural algorithms based on CNN, using parallel computing of the open Hadoop and Spark frameworks, to form complex forecasts in the field of economics and management, including forecasting the company's net profit. It seems appropriate to consider the issues raised through the prism of the neuroecosystem model of the Industry 5.0 concept, which, according to Babkin A. et al. will allow us to set the task of implementing the systems of global meta-system strategic development of cognitive production and industry [15]. The author Meyer C. identified the main directions for the development of hybrid cyber-physical, cognitive-social systems in line with the development of technologies for the new technological redistribution Industry 5.0 [16].

A team of authors led by Klachek P. M. proposed a very promising approach, which has an interdisciplinary basis, since it is on the border of the following areas: hybrid intelligent systems, synergetic artificial intelligence, neuro- and psychophysiology, philosophy, cybernetics, economic and mathematical modeling, etc. [17]. In [18–20], the authors considered in detail the main features and formal foundations of complex weakly formalizable multicomponent economic systems (MES) and the corresponding difficult to formalize production and economic problems.

The main goal of the research is the development of effective methods and the search for applied tools for improving the efficiency of intelligent information processing and management systems, namely, computer decision support systems that are used when performing hard-to-formalize production and economic tasks based on functional hybrid intelligent decision-making systems (FHIDMS). Such systems, as practice has shown, are able to successfully solve complex hard-to-formalizable production and economic problems and develop solutions of appropriate quality in various subject areas. Among the important areas that deserve attention for future research, the following topics should be noted. Also of note is a study by Greg Buchak and his colleagues to identify the drivers driving fintech adoption [21]. To ensure sufficient detail, it seems appropriate to correlate the obtained results with similar works and directions of other studies. It is also advisable to offer options that reflect the results obtained in previous studies.

6   Conclusions

Based on the results obtained by the HAM, Caustic should change the list of contractors, terminate contracts with contractors with a low rating of experts and conclude contracts with contractors with a high rating. This will reduce the volume of annual price increases. The expected economic effect will be 15,180 thousand rubles. Digital-Bot for collecting information about construction contractors, their ranking and selection based on the hierarchical analysis method (HAM), was successfully formed.

The result of the study was the predicted values of the net profit of the Caustic company based on the use of a neural network Deep learning model "Random Forest".

The hypothesis put forward is proved that using the Random Forest deep learning model, you can get the predicted value of the company's net profit. Using the parameters of the RF model, based on the values of factor signs for 2022, made it possible to obtain the forecast value of the current sign–the value of the net profit of Caustic for the next year 3,749,506.9 thousand rubles.

Data analysis allows us to conclude the average level of forecast accuracy. So the level of the average absolute error of the model was 714,688.25 thousand rubles. in the case where the share of the test sample in the dataset is 0.30. At the same time, the MAE error increased to 1,131,354.33 thousand rubles, or 1.58 times, if the ratio of the test sample is reduced to 0.20.

This study provides an increment of scientific knowledge that allows closing the scientific gap in terms of identifying and evaluating the factors affecting the net profit of a company in the real sector of the economy in modern conditions. Further research can be directed to the use of more advanced tools–convolutional neural networks with high performance.

The priority direction for the development of enterprises in the real sector of the economy is the wider use of artificial intelligence systems. In addition, a promising direction is cyber-physical systems based on Industry 5.0 technologies.

References

1. Caustik: accounting and financial analysis https://www.audit-it.ru/buh_otchet/3448003962_ao-kaustik. Accessed 20 Aug 2023
2. Hierarchy Analysis Method https://ru.wikipedia.org/wiki/%D0%9C%D0%B5%D1%82%D0%BE%D0%B4_%D0%B0%D0%BD%D0%B0%D0%BB%D0%B8%D0%B7%D0%B0_%D0%B8%D0%B5%D1%80%D0%B0%D1%85%D0%B8%D0%B9. Accessed 20 Aug 2023
3. Kvartina VV, Inshakova EI (2019) Formation of a system of relationships between the participants of the investment process in construction. In the collection: Management of socio-economic development of regions: problems and ways to solve them. In: Collection of scientific chapters of the 9th international scientific and practical conference, vol 3, pp 277–289
4. Kovalchuk SI (2019) Formation of a system of relationships between participants in the investment process in the construction of the region (based on the materials of the Smolensk region). In the collection: Theoretical and applied aspects of scientific research. In: Collection of chapters based on materials of the IV annual international scientific and practical conference, pp. 46–52
5. Bataev AV, Gorovoy AA, Denis Z (2019) Comparative analysis of the use of neural network technology in the world and Russia. In: Proceedings of the 33rd international business information management association conference, vol 2, pp 988–995. https://www.scopus.com/record/display.url?eid=2-s2.0-85073376796&origin=resultslist. https://doi.org/10.34190/ECIE.19.165
6. Shokhnekh A, Lomakin N, Glushchenko A, Sazonov S, Kovalenko O, Kosobokova E (2019) Digital neural network for managing financial risk in business due to real options in the financial and economic system. In: Conference: proceedings of the international scientific-practical conference "Business cooperation as a resource of sustainable economic development and investment attraction" (ISPCBC 2019). https://doi.org/10.2991/ispcbc-19.2019.138
7. Lin II, Zhou D, Liu W, Bian J (2021) Deep risk model: a deep learning solution for mining latent risk factors to improve covariance matrix estimation. In: ICAIF'21. November 3–5. Virtual Event. USA. https://arxiv.org/format/2107.05201. Accessed 23 June 2023
8. Nakagawa K, Noma S, Abe M (2020) RM-CVaR: regularized multiple β-CVaR portfolio. IJCAI-PRICAI special track AI in FinTech. https://doi.org/10.48550/arXiv.2004.13347
9. D'Hondt C, De Winne R, Ghysels E (2019) Steve, Raymond artificial intelligence alter egos: Who benefits from Robo-investing? Portfolio Management (q-fin.PM); Econometrics (econ.EM); Statistical Finance (q-fin.ST). https://doi.org/10.48550/arXiv.1907.03370. Accessed 23 June 2023

10. Krzysztof R, Piotr B, Jaglarz P, Fabien G, Albert C, Piotr C (2022) RiskNet: neural risk assessment in networks of unreliable resources. ACM classes: I.2; C.2//Submitted 28 January, 2022; originally announced January (2022). https://doi.org/10.48550/arXiv.2201.12263

11. Clarkson J, Cucuringu M, Elliott A, Reinert G (2022) DAMNETS: a deep autoregressive model for generating Markovian network time series. Statistical Finance. https://doi.org/10. 48550/arXiv.2203.15009. Accessed 23 June 2023

12. Bingyan H (2023) Distributionally robust risk evaluation with causality constraint and structural information. Mathematical Finance. 20 Mar. https://doi.org/10.48550/arXiv.2203.10571. Accessed 23 June 2023

13. Neufeld A, Sester J, Yin D (2023) Detecting data-driven robust statistical arbitrage strategies with deep neural networks. Mach Learn. arXiv:2203.03179. https://doi.org/10.48550/arXiv. 2203.03179. Accessed 23 June 2023

14. Lomakin N, Sazonov S, Polianskaia A, Lukyanov G, Gorbunova A (2020) Artificial intelligence system for processing big data to determine the value of innovative products in a digital economy. In: Communications in computer and information science this link is disabled, 2020, 1273 CCIS, pp 19–37

15. Babkin AV, Fedorov AA, Lieberman IV, Klacek PM (2021) Industry 5.0: concept, formation and development. Russian J Ind Econ 14(4):375–395

16. Meyer C (2021) Model risk in credit portfolio models. Risk Manag. arXiv:2111.14631v1. https://doi.org/10.48550/arXiv.2111.14631. Accessed 23 June 2023

17. Klachek PM. Babkin AV, Liberman IV (2019) Functional hybrid intelligent decision-making system for hard-to-formalize production and economic tasks in the digital economy. Scientific and technical statements of SPbSPU. Econ Sci 12(1):21–32

18. Gavrilov AV (2003) Hybrid intelligent systems. Publishing House of NSTU, Novosibirsk

19. Klachek PM, Polupan KL, Koryagin SI, Liberman IV (2018) Hybrid computational intelligence. Fundamentals of theory and technology for creating applied systems. Izd-vo BFU im. I. Kant, Kaliningrad

20. Kolesnikov AV (2001) Hybrid intelligent systems: theory and development technology. Publishing house of St. Petersburg STU, St. Petersburg

21. Buchak G, Matvos G, Piskorsk T, Seru A (2019) Fintech, regulatory arbitrage, and the rise of shadow banks. J Financ Econ 133(1):18–41. https://doi.org/10.1016/j.jfineco.2019.01.006

Digital Transformation of Heat Supply and Unified Heat Supply Organizations Based on the Introduction of Digital Control Elements

Valeriya Glazkova (ID)

Abstract Decrease in the efficiency of the heat supply in Russia due to the high level of deterioration and insufficient modernization of the system cause the necessity to find the ways for its development. As a result of the analysis of the practice of implementing heat supply digital and information solutions, performed within research, an increase in the efficiency and reliability of heat supply organizations and systems through their use was noted. The importance of using a systematic approach to the digital transformation of heat supply, which is implemented through unified heat supply organizations, is emphasized, since they are responsible for monitoring the main components of the heat supply system functioning, including the processes of its digitalization. Therefore, before planning and implementing the digitalization of the heat supply system, it is necessary to transform existing approaches to the organization and implementation of unified heat supply organizations business processes. For such processes digital transformation is considered as a form of managerial, organizational and technological changes which contribute to improving the efficiency of both the organizations themselves and heat supply systems in the territories assigned to them, as well as influencing the development of digitalization of the heat supply industry in general. This provision determined the purpose of the research, which is to develop a model for implementing the digital transformation of unified heat supply organizations. By summarizing individual elements and conditions for the digital transformation of organizations, the author proposes a model for implementing digital transformation of unified heat supply organizations. This model is detailed at the corresponding stages, including description of the tools used at these stages, as well as indicators and directions reflecting the effectiveness of the digital transformation of the organizations under consideration. Obtained results reflect a consistent approach to the implementation of unified heat supply organizations digital transformation, including an analysis of the current state and prospects of digital transformation, definition of its main goals and indicators, development of digital transformation measures, directions for their implementation, evaluation and

V. Glazkova (✉)
National Research Moscow State University of Civil Engineering (NRU MGSU), 26
Yaroslavskoye Highway, Moscow, Russia
e-mail: leram86@mail.ru

© The Author(s), under exclusive license to Springer Nature Switzerland AG 2023   105
A. Bencsik and A. Kulachinskaya (eds.), *Digital Transformation: What is the Company of Today?*, Lecture Notes in Networks and Systems 805,
https://doi.org/10.1007/978-3-031-46594-9_7

control. Its' successful implementation contributes both to increasing efficiency of unified heat supply organizations activities, as well as to improving reliability and reducing accidents in heat supply systems.

Keywords Digital transformation · Information technologies · Digital control · Sustainable development · Heat supply · Unified heat supply organizations · Heat supply systems

1 Introduction

Heat supply, as an integral part of the fuel and energy complex of the Russian Federation, highly contributes to the growth of the national economy, also has a significant socio-economic impact and affects the sustainable development of the territories. Researchers of the heat supply functioning and development underline a number of problem areas and negative trends that have developed in the area under consideration. It includes a high level of fixed assets depreciation of the heat supply system of the Russian Federation [1], its' inadequate renewal and modernization [2], large losses of thermal energy during transportation, high costs and accident rate on the heating systems [3].

According to Terentyeva [4], Verstina et al. [5] and others, these processes have led to a crisis state of heat supply in Russia. Industry experts note the high deterioration of the heat supply system. Ultimately, it leads to low efficiency and a drop in the level of reliability of heat supply system (Figs. 1 and 2) [2.6]: by 2020, heat losses in heat systems have increased by 2.51 times compared to 1995. The number of heat systems in need of replacement is growing every year and by now already exceeds 30% of all heat and steam systems. In this regard, becomes relevant a more detailed research of the heat supply current state and identification of possible areas for its development.

Obsolete capacities are characterized by low reliability and high repair costs. Long-term downtimes lead to an increase in the cost of heat production [2, 8]. Such unreasonable costs lead to unprofitability of the heat supply industry, which has been observed since 2005 (Fig. 3).

According to the provisions of the Federal Law of July 27, 2010 N190-FZ "On Heat Supply" [10], unified heat supply organizations (hereinafter referred to as UHSO) should integrate the functions of management, control and distribution of information regarding the operation of the heat power engineering industry production capacities and district heating systems. At the same time, a number of researchers agree that functioning of the UHSO should be facilitated by the digitalization of control systems in the industry [5, 11–13]. And development of a new digital format for the organizational integration of a group of heat and power engineering organizations by streamlining the renewal and modernization of production assets, based on the implementation of UHSO, will overcome the problem of the low level of

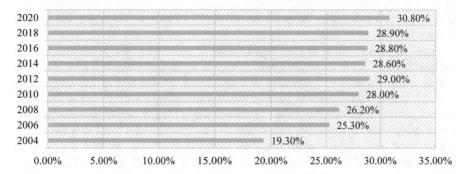

Fig. 1 The share of domestic heat and right side systems in two-pipe terms that need to be replaced in the total volume of domestic heat and steam systems, % (compiled by the author on the basis of Rosstat data)

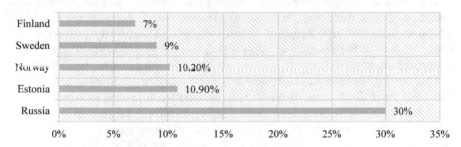

Fig. 2 The share of losses in the total volume of generated heat energy as of 2018 in countries with a central heating system (compiled by the author on the basis of the research of Gavrilenko and Khakimov [7])

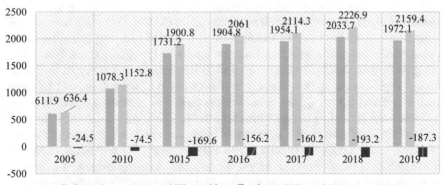

Fig. 3 Dynamics of heat power engineering revenues and costs and of district heating for the period 2005–2019 (compiled by the author based on data from the Russian Ministry of Energy [9])

the industry controllability, increase the efficiency of heat supply organizations and reduce their losses [14–16].

As of 2017, only 65% of heat supply organizations had used basic information technologies, which include the use of computers, the Internet, local area networks and automated resource accounting systems. It was considered that the legislative measures, adopted in the same year, aimed at stimulating development of the information society and digital economy in the country [17, 18], would make it possible to intensify work on the intellectualization of the Russian fuel and energy complex (FEC). And heat supply system is a component of FEC [19]. However, according to the results of 2021, in the total revenue of more than 4.7 thousand heat supply organizations, which amounted to about 2 trillion rubles, the amount of costs for the implementation of information technologies amounted to just under 20 billion rubles [16], which indicates a low rate of heat supply digitalization.

In accordance with global trends, digitalization is an integral process for the development of both housing and communal services and the fuel and energy complex: it is enough to recall the concepts of "smart society" and "smart city", which predetermine the need for widespread introduction of information technologies in the designated areas.

Considering the international experience of heat supply digitalization, it should be noted that the situation in the world with this process is controversial, and sometimes contrasting: Finland is recognized as the leading country in this regard, which has introduced almost universal digitalization of the heat supply system both in large cities and in small towns. Similar situation is typical for the US, UK, Japan, Germany, Canada and Italy, and other advanced economies. However, it is worth noting that projects for the heat supply digitalization, which include digital accounting, use of geographic information systems, call center automation, dispatching, information support for network connecting, payment processes, etc., are implemented in these countries largely by private organizations, operating in the energy and heat market.

Speaking about the place of Russia in the process of heat supply digitalization, researchers of this issue note that domestic heat supply organizations, both public and private, undoubtedly realized the need to increase the efficiency of their work through digital systems. However, the implementation of projects is just beginning, and there are only a few organizations already completed implementation [20, 21]. At the same time, according to the estimates of the HSE Institute for Statistic Research and Economics of Knowledge, the demand of the domestic fuel and energy complex for digital technologies can grow up to 13.5 times: from 30.7 billion rubles in 2020 to 413.8 billion rubles in 2030. Moreover, it looks even more relevant due to the refusal to supply and work on the Russian market of a number of foreign vendors, which will affect the increase in demand specifically for domestic digital solutions and products [22].

2 Methods and Methodology

Implementation of digital technologies is accompanied by the digital transformation of UHSO, with its further contribution to the improvement of the processes for managing development of the organizations in question and heat supply systems assigned to them. In this respect, it becomes necessary to consider the concept of "digital transformation", which, according to the research, does not currently have a single interpretation. This term was mentioned first in relation to the characteristics of a significant change in the external and internal organization environment at the end of the twentieth century. Further, the analysts of the Boston Consulting Group defined digital transformation as "the fullest use of the potential of digital technologies in all aspects of business" [23]. A similar point of view was expressed by many Russian researchers (Garifullin and Zyabrikov [24], Ovchinnikova et al. [25] and others), considering digital transformation in relation to the micro-level of management, They meant by it directly the digital control system of the organization as an element of digital transformation, which uses modern digital technologies as a tool [25].

However, according to a number of researchers' opinion digital transformation should be considered not only at the micro-level, but should as well be extended beyond the organization. Zaichenko et al. [26] determine digital transformation as a "step in the development of the global economic system". Vyugina [27] also notes in her works the scale and scope of the digital technologies use in the course of digital transformation. The Russian Guidelines for the Digital Transformation of State Corporations and Companies with State Participation [28] consider the possibility of applying the term "digital transformation" both to organizations and to the industry, determining that the provisions of this official document can be used as recommendations for any organization and industry.

As to the subject of this research–unified heat supply organizations and heat supply systems assigned to them, it should be specified that the process of digital transformation for UHSO should imply managerial, organizational, technological changes that contribute to improving the efficiency of the organizations in question, heat supply systems assigned to their territories, as well as affecting the development of heat supply industry digitalization in general. Since successfully implemented measures for the digital transformation of one heat supply entity can be scaled for another entity. In this regard, further in the research it is advisable to define the main areas for the implementation of digital technologies in heat supply systems and organizations and to develop a common approach to the digital transformation of UHSO.

Currently in Russia, computerization of heat supply systems has become widespread, represented by developed software and hardware systems. Its' basic element is automated control systems that perform input–output, processing, archiving, and exchange via local and remote interfaces [29]. Experts underline that when using automated dispatch control systems, heat savings range from 25 to 40% due to the rational use of resources and optimized control of heat supply systems. According to the results of the research by V.I. Solovyov, who analyzed the operation of the district heating system of the city of Novosibirsk, by reducing losses only at the stages of transportation, distribution and consumption to the "Scandinavian" level, it is possible to achieve the heat saving potential and increase in heat efficiency for Novosibirsk heating systems in the amount of at least 27% [19]. Thus, the automation of heat supply systems, as a step towards its digitalization, has a positive effect both on the efficiency of managing heat supply systems and on their performance of environment safety properties in view of the fact that it allows you to save and optimally use resources, reducing their consumption. Therefore, it contributes to reduce emissions to the environment and reduce the overall level of pollution.

In order to expand the use of digitalization elements in heat supply organizations and systems, the Ministry of Energy of the Russian Federation, with the participation of fuel and energy companies, formed the initiative "Digital Energy" [30]. The purpose of its implementation is digital transformation of the fuel and energy sector, taking into account the priorities of the national program "Digital Economy of the Russian Federation", aimed at transforming the domestic energy infrastructure through the introduction of digital technologies and platform solutions to improve its efficiency and safety [31]. According to A.V. Novak, the key organizational task of the project is to build a system for coordinating the digital transformation of the Russian fuel and energy complex [6].

The Digital Energy initiative also concerns the heat supply digital transformation, which includes modernization based on the introduction of digital and information technologies not only of facilities for the production, transmission and distribution of heat, but also the digitalization of heat supply system management processes. With regard to the development of the UHSO, effective implementation of the heat supply digital transformation will improve the efficiency of operations through the introduction of modern technologies for collecting and processing data, increasing efficiency of the heat power infrastructure, rational use of resources and, as a result, fulfilling the property of environmental safety, increasing reliability and reducing accidents in heat supply, a decrease in the final price of heat and the emergence of demanded services to meet the needs of consumers. In other words, it will satisfy the interests of all parties interested in the development of UHSO and heat supply system (Table 1).

Digitalization of UHSO and the heat supply system assigned to it is not an aim in itself, but is one of the components of the development of domestic heat supply, which makes it possible to increase the efficiency of its functioning. Based on the data presented in Table 1, it is safe to say that all UHSO stakeholders will benefit from the digitalization of heat supply. If we translate the effects of the introduction of digital technologies into heat supply into numbers, then, according to researcher E. Goncharov, "by reducing the number of accidents and defects by 1% per 500 km of heating systems in a small city, heat supply organizations save up to 100–150 million rubles a year on repairs. It is quite realistic to achieve such a result in practice through the use of digital systems in the work of heat supply organizations" [21].

As part of the implementation of the Digital Energy initiative, in 2022, a pilot project Digital Heat Supply was launched (developer by JSC Rusatom Infrastructure

Table 1 List of positive effects for the main stakeholders in the process of introducing digital technologies in UHSO and heat supply systems (compiled by the author based on materials on the practical application of digital technologies in heat supply from open sources [32])

Stakeholder	Positive effect of digital technology introduction
Public authorities	1. Decreased dissatisfaction of people with the quality of services provided, which is indirectly focused on the work of state and municipal governments 2. Digitalization of the heating system operation will make it possible to quickly and transparently respond to residents' questions regarding heat supply services 3. Reducing service costs and, as a result, preventing the growth of tariffs 4. Reducing the cost of repairing and replacing equipment and pipe sections through the introduction of digital technologies saves budget funds 5. Digitization of heat supply will make it possible to plan investment programs more efficiently
Supervisory Authorities	1. As a result of automation of heat supply systems, there is a reduction in heat losses and resource saving, which affects the reduction of emissions into the environment
Shareholders	1. Predictive analytics allows you to plan the expenses of organizations for the year ahead more accurately, optimize resource costs 2. Digitalization of interaction with buyers will ensure transparency and efficiency of fundraising from the management company and the population, which will not only affect the profits of heat supply organizations, but will also, stop the rapid growth of debt for utilities 3. Digitization of heat supply will allow more efficient planning of investment programs
Investors	1. Predictive analytics allows you to plan the expenses of organizations for the year ahead more accurately, optimize resource costs 2. Digitization of heat supply will allow more efficient planning of investment programs
Consumers	1. Forecasting the places of damage in systems will lead to a decrease in the number of accidents and an increase in the reliability of heat supply 2. Increasing awareness and quality of services through the introduction of services for online application for the conclusion of contracts, the exchange of payment documents and online payment for heat 3. Reducing service costs and, as a result, preventing the growth of tariffs 4. Due to the ability to automatically regulate the flow of coolant thanks to weather-dependent technology, the consumer will receive a comfortable temperature regime
Local communities and future generations	As a result of heat supply systems automation, there is a reduction in heat losses, an increase in the energy efficiency of systems by reducing heat losses and excessive fuel generation at heat sources, which affects the reduction of emissions into the environment
Public organizations and ecological movements	1. As a result of heat supply systems automation, there is a reduction in heat losses, an increase in the energy efficiency of networks by reducing heat losses and excessive fuel generation at heat sources, which affects the reduction of emissions into the environment 2. Due to the ability to automatically regulate the flow of coolant thanks to weather-compensated technology, which allows you to keep accurate records of resource consumption, you benefit from energy-efficient measures

(continued)

Table 1 (continued)

Stakeholder	Positive effect of digital technology introduction
Organization employees	1. Predictive analytics allows you to plan the expenses of organizations for the year ahead more accurately 2 Due to the complete picture of work efficiency, the management of the enterprise can introduce various motivation and incentive schemes for staff, which means that the income of employees will increase
Heat supply organizations	1. Predictive analytics allows you to plan the expenses of organizations for the year ahead more accurately, optimize resource costs 2. Predicting fault locations in systems will lead to a reduction of the amount of accidents and to an improvement in the functioning of organizations, as well as to a reduction in response time and repair of damage 3. Automated thermal energy accounting system allows you to collect and store automatically readings from metering stations necessary for the analysis and control of heat supply parameters 4. Improving the quality of operational dispatch control 5. Due to the use of digital twins, artificial intelligence, automation of the accounting and control system, which lead to the rapid identification of damage points in networks, the number of damaged heating networks required repair is reduced 6. Digitization of interaction with buyers will ensure transparency and efficiency of fundraising from the management company and from the population, which will not only affect the profits of heat supply organizations, but will also stop the rapid growth of debt for utilities
Construction and installation organizations	Due to the use of digital twins, artificial intelligence, automation of accounting and control systems, which lead to the rapid identification of damage points in networks, the number of damaged heating networks that require repair is reduced
Design organizations	Digitalization of processes for approval of projects for the construction and reconstruction of heat supply facilities will reduce the financial and time costs for project approval

Solutions), which is a fully import-independent integrated solution for digital transformation and automation of the main business processes of resource supply organizations. The product is based on the use of technologies such as digital twins (digital twins of engineering infrastructure, generation facilities, technological connections of equipment and their various variations are created to achieve maximum efficiency), artificial intelligence (to identify network sections where an accident can occur) and big data technologies (creation of a common information space for resource management and digital control over the implementation of measurable indicators).

The Digital Heat Supply system can be implemented as a complete set of solutions for heat supply organizations or as a module to automate individual business processes in an organization (such as sales, repair procedures, transport, labor protection, etc.). With the help of planning, monitoring and control tools, the modules allow you to automate processes in all major areas of activity of heat supply organizations, from heat generation to interaction with end consumers [33]. The efficiency of using of Digital Heat Supply modules is reflected in reduction in heat losses (by 20%), reduction in the number of breakdowns (by 25%), an increase in the efficiency of identifying emergencies (by 90%), increase in labor productivity (by 60%), and also increase in energy efficiency (by 15%) [34].

Efficient operation of modules of the "Digital Heat Supply" system allows increasing the reliability of heat supply, as well as reducing resource consumption, thereby affecting the performance of the environmental safety. As for the processes of heat supply system managing such a digital product, contributes to the efficiency of heat supply organizations business processes to increase, including unified heat supply organizations. It has positive effect on the speed and accuracy of making management decisions, ultimately affecting increasing efficiency of UHSO and heat supply system assigned to it. In addition, one of the system's advantages is settled by a developer opportunity to provide authorities with information on the operation of heat supply organizations in the form of reports. For the moment, the practical significance of the "Digital Heat Supply" system is confirmed by its use in Glazov city (Udmurt Republic), Lipetsk and Voronezh [33].

The practice of applying digital twins in the process of managing the heat supply systems operation has already begun to pay off. Thus, in 2021, an automated heat supply control system was put into operation in Yekaterinburg, as well as "digital twin" of the heat supply system based on it. As a result, the number of damages to heat systems reduced by 10%, heat losses–by 1.5% [34]. Experts point out that due to the application of digital twins in heat supply it is possible to achieve a 40% decrease of excess losses and save up to 15% of fuel burned.

Another area of using digital technologies in heat supply systems is application of robots for the internal condition of the heating line diagnostics. A remotely controlled robot, moving, detects existing defects: cracks, chipping, dents, corrosion, etc., defines location of damage, and measures the wall thickness around the entire perimeter of the pipe. In addition, the computer program is reporting on the criticality of the defect detected and on the allowable period of the pipe continue operating. When fixing that no further use is permitted, it contributes to the choice of the required pipe repair method [19]. Thus, the application of the considered digital solutions leads to decrease in the accident rate and to increase in the reliability of the heat supply system.

Speaking about the benefits of heat supply digitalization, we should raise the issues of improving management processes of heat supply organizations, including unified heat supply organizations. Since this process stimulates the development of UHSO by business processes automation: preparation of initial and operational reports on heat supply organization and system in the territory assigned to it; personnel systems development; use of digital technologies in the field of labor safety and health protection; development of information and personal data security, and other areas that contribute to the sustainable development of UHSO.

Areas of heat supply digitalization considered above, make it obvious that this process covers the main components of the heat supply system: the source of heat energy, the heating system itself and the consumer of heat energy. Supervision of the operation of these components is entrusted to UHSO. Therefore, within the framework of the unified heat supply organizations activities, before planning and implementing the digitalization of the heat supply system, it is necessary to transform existing approaches to the organization and implementation of business processes related to the development of UHSO. As well as to perform infrastructural, organizational and technical transformations throughout the entire chain of the technological process "generation–transportation–heat energy consumption" [19].

However, some experts point out that with all the positive aspects of the digital technologies introduction in the management processes of UHSO and heat supply system, practically it is too early to talk about heat supply digitalization until the issues of high level of heat supply systems deterioration, poor management and underfunding of the industry have been resolved. It is because of the fact that introduction of information and a number of problematic issues that will have to be faced as well accompanies digital technologies. Yarmiev and Chernikova [29] distinguish the following as difficulties in the implementation of digital and information technologies in heat supply:

– scale of the Russian fuel and energy complex, influencing the need to develop a large amount of solutions, technologies, and means for its widespread implementation with continuing underfunding of the industry;
– lack of understanding and underestimation of the modern digital and information technologies opportunities and effectiveness of its implementation by the leaders of heat supply organizations;
– lack of competent specialists in information and digital technologies, as well as domestic developments in this area.

Researchers L.V. Gurianova, A.Yu. Ugrevatov and A.N. Kulikov, completing the challenging areas of heat supply digitalization with the thesis that the use of a significant number of obsolete equipment, which is also of high level of wear and tear, also complicates implementation of digital technologies. These authors in their research propose first to deal with the reasons that led to the prevailing heat supply state, to eliminate it in the most cost-effective way when there is insufficient investment in the industry, and then consider digitalization of the industry [22].

E. Goncharov observes that for organizations responsible for the implementation of heat supply systems digital transformation, the main challenge is to find a balance between current operating and upgrading costs and costs for implementation of digital technologies, in other words, costs that have a strategic perspective, and no impact momentary. This challenge is compound by the lack of financial resources and lack of motivation to implement digital projects among the leaders of heat supply organizations [21].

According to O.P. Ovchinnikova, M.M. Kharlamova, T.V. Kokuytseva, digital transformation of organization will require compliance with the following conditions: high quality and readiness of all information and communication systems, synchronization of data and information for decision-making at all levels of management, changes in the organization-operating model and organizational structure [25].

3 Research Results

It is obvious that compliance with the conditions of organizations digital transformation as all its information and communication systems are ready for change, the need to transform the operating model and revise the organizational structure [25], during the UHSO digital transformation process, will entail changes in the management processes of the organizations in question. In this regard, we will form a model for implementing UHSO digital transformation, and will perform this process in a graphical way on Fig. 4.

At the first stage of digital transformation, UHSO needs to assess the current state of the organization in terms of its readiness for digital transformation, as well as the state of the macro- and meso-environment in terms of the current trends in digital development in the industry, as well as the opportunities and threats that the macro environment provides. The resulting part of this stage is identification of promising areas for UHSO digital transformation. Description of the tools used at this stage is performed in Table 2.

Based on the identification of promising areas for UHSO digital transformation, a system of goals and indicators of digital transformation is formed at the next stage. To do this, the target business model of UHSO is determined. It includes description of the organization's business processes in the context of digital transformation. Forming the digital business models is one of the main tools for the digital transformation of an organization. Its flexibility allows quickly changing key parameters of UHSO business processes and responding faster to changing external conditions, thereby increasing the efficiency of both UHSO and heat supply system assigned to it. After building a business model, digital transformation goals are set. Goals are detailed into tasks that should necessarily take into account the interests of the main stakeholders in the process of implementation of UHSO digital technologies (Table 1), and include an assessment of the contribution of digital transformation to improving the performance of UHSO and heat supply system.

For the identified goals and objectives of UHSO digital transformation, it is necessary to determine the planning horizons (which of them are long-term, medium-term and short-term), as well as volume and sources of resources needed to achieve them. Based on the current indicators of UHSO functioning, as well as taking into account the goals of the digital transformation of the organization and the timing of their achievement, a system of indicators is formed. It reflects the effectiveness of UHSO digital transformation, which is used in the future to control and monitor the process under consideration. UHSO digital transformation performance indicators depend largely on the goals set for the digital transformation of organizations. The goals, in turn, are determined by the characteristics of the current state of the organizations in question, as well as the opportunities and threats of the external environment. However, the Guidelines for the Digital Transformation of Public Corporations and Companies with State Participation offer a list of general indicators that can be used to assess the effectiveness of the digital transformation of organizations. In our opinion, these indicators are also applicable to determine the effectiveness of UHSO digital

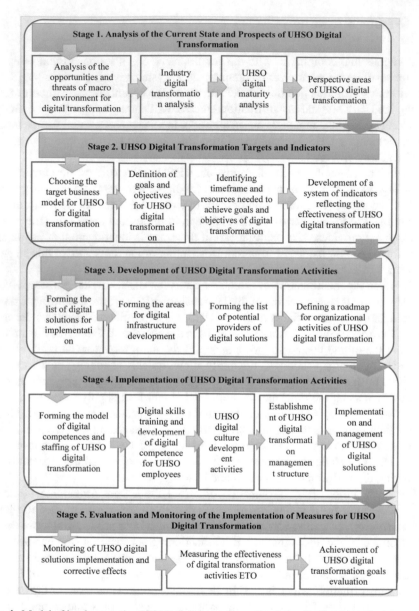

Fig. 4 Model of implementation UHSO digital transformation (compiled by the author)

transformation (Fig. 5). The list of indicators, shown in Fig. 5, can be supplemented with performance indicators depending on the goals and objectives of UHSO digital transformation.

At the next stage, based on the goals and objectives of digital transformation, measures for their implementation are developed. They start with a list of digital

Table 2 Tools used at the stage of analysis of the current state and prospects of UHSO digital transformation (designed by the author)

Direction for analysis	Content of the analysis	Applied tools
Analysis of the opportunities and threats of macro environment for digital transformation	The current state and potential of the macro-environment for UHSO digital transformation is assessed in relation to the best domestic and foreign practices in various sectors of the national economy. As well as the quality of external conditions for the heat supply industry digital transformation, that could trigger digital development or, conversely, create additional barriers (conditions and financing instruments, state support measures, national programs, etc.)	PEST-analysis, SWOT-analysis, expert evaluation
Industry digital transformation analysis	The change in the behavior of heat supply organizations in the implementation of new business models, digital platforms, information technologies, changes in consumer behavior associated with the introduction of digital services, etc. are assessed	Analysis of 5 competitive forces of Michael Porter, expert evaluation, analysis of the practice of implementing digital and information systems in the heat supply organizations operation process, including UHSO
UHSO digital maturity analysis	The current level of UHSO digital and information technologies usage is assessed, as well as promising key areas for UHSO transformation and existence of an enabling environment in UHSO	SWOT-analysis, SNW-analysis, methodology for evaluating UHSO innovation readiness, methodologies for assessing the digital maturity of industrial enterprises [28, 35]

solutions that can be implemented in UHSO. A typical approach to compiling a list of digital solutions should be based on the criteria of conformity of UHSO readiness to implement the solution, compliance with objectives of UHSO digital transformation, as well as the greatest impact of the digital solution. For the selected solutions for UHSO digital transformation, appropriate digital infrastructure development needs to be identified, taking into account the results of its current state analysis. Further determines whether the own digital solutions will be used, or external providers of digital solution are preferable. When choosing in favor of the latter, a list of bona fide potential digital solutions suppliers is established. When compiling a roadmap of organizational activities of UHSO digital transformation, a list of organizational activities within the framework of digital transformation is provided. It should include a brief description of each activity, as well as planned changes in the organizational

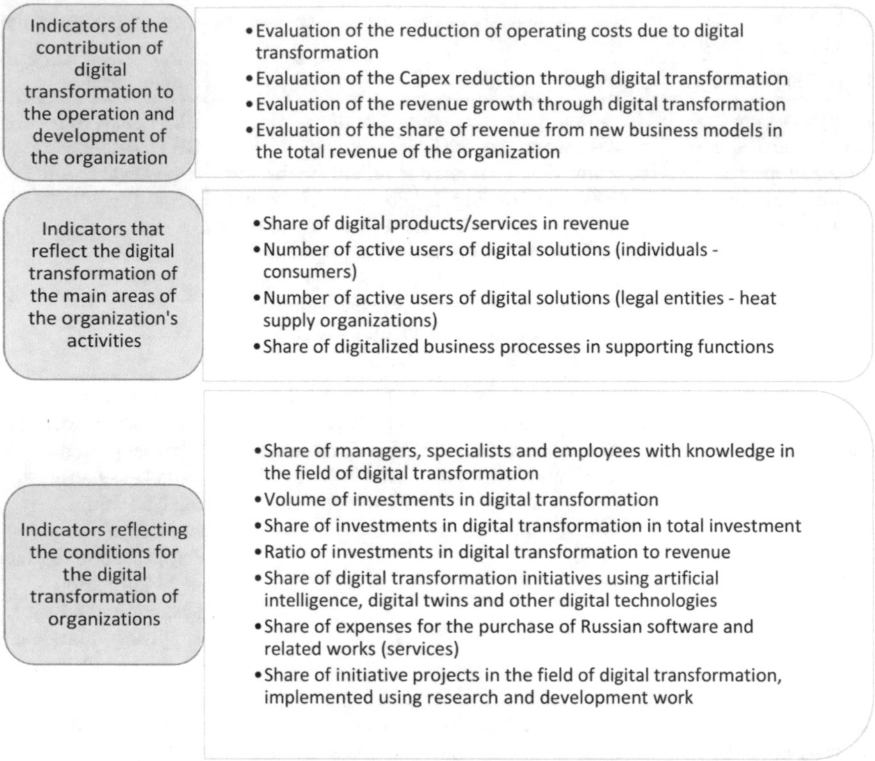

Fig. 5 The list of indicators for evaluating the effectiveness of the digital transformation of an organization (compiled by the author on the basis of the Guidelines for the digital transformation of state corporations and companies with state participation) [28]

structure of UHSO in connection with digital transformation. For example, development of digital competence centers, of the head position for digital transformation etc. As well, description of the competencies, job responsibilities and powers for employees involved in UHSO digital transformation process should be included.

Implementation of measures for the digital transformation of UHSO should start with the development of a digital competence and digital staffing model for the UHSO digital transformation, including an assessment of staffing need based on the list of digital solutions for implementation formed earlier. In addition, it is necessary to conduct digital skills training for UHSO employees involved in digital transformation processes. Measures to develop a digital culture include initiatives such as the implementation of customer-oriented approaches to work, operating practices in the context of constantly changing requirements, feedback services for UHSO employees, etc.

The digital transformation management structure includes establishment of special units (committees, supervisory boards, digital transformation office, etc.), description of its place in the overall UHSO management system, as well as description of the functions, powers and composition of the units. Experts point out that in the context of digital transformation, the most effective is the flat structure of the organization, which implies greater independence of individual departments of the organization, since through the introduction of digital technologies, more flexible ways of developing and manufacturing products will become possible (for example, through the use of virtual and augmented reality technologies, creating digital twins, as well as through 3D modeling), bringing them to the market (big data analytics), optimizing the supply chain [25].

The direct implementation of digital and information products into production and management processes of UHSO, as well as its management, is also carried out at the implementation stage. This process involves digitalization of UHSO operating model that contribute to optimize business processes of the organization.

At the final stage of UHSO digital transformation, it is necessary to monitor the implementation of digital solutions in UHSO, depending on its results, to perform corrective actions at stages 2 and 3 of the model considered (Fig. 4). In addition, it is necessary to evaluate all types of effects from the digital transformation of UHSO, including both typical indicators performed in Fig. 5, as well as indicators that reflect the achievement of goals of the UHSO main stakeholders, listed in Table 1, including indicators related to the social and environmental effectiveness of UHSO digital transformation.

4   Conclusion

Thus, as a result of digital transformation, the operation and development of UHSO occurs with usage of information and digital technologies, which are the basis of the organization's business processes. It should be noted that the digital transformation of UHSO will affect not only the internal environment, but also the "external contour" will be changed. That is, the nature of interaction with external stakeholders will be changed, since management decisions will be based on "digital analytics" [25].

As for the implementation of digital transformation in practice, a positive point is that "digitalization in heat supply, as a rule, is carried out "from the bottom up" (heat grids are equipped with devices and sensors for data processing, from which an information system of a heat supply organization is developed). It gives an opportunity to scale up the system gradually" [1, 3]. However, implementation of these activities will require the addition of new functions to the processes of managing the heat supply system, and therefore it is important to initially use a systematic approach to the heat supply digital transformation, implementing it through unified heat supply organizations.

Thus, the analysis of digital technologies in heat supply use, carried out as a part of this research, made it possible to determine its significant role not only in improving the efficiency of industrial and technical heat supply systems, but also in the processes of managing these systems. The model for the implementation of

UHSO digital transformation, proposed by the author, is aimed at the sustainable development of the organizations in question and, as a result, the heat supply systems assigned to them, which will improve their reliability and efficiency.

At the same time, speaking about a systematic approach to the digital transformation of heat supply, it is necessary to raise the issue of a unified approach to the use of information and digital technologies in promising researches. The Russian experience of digital transformation of heat supply considered in the given research reflects that heat supply organizations work in this direction separately, often not taking into account the experience and competencies of each other. As a result, the same tasks can be solved in different ways using separate parts of complex solutions. In this regard, there is a need to create digital and information platform solutions for heat supply, which in turn will require the formation of data representation and standardization of processes associated with its use, which is a promising management task in the framework of achieving the goal of effective digital transformation of heat supply.

References

1. Tsuverkalova OF (2020) Analysis of the current state and trends in the development of the heat supply industry in the Russian Federation. Bull Altai Acad Econ Law 2020(11–3):554–559
2. Stennikov V, Penkovskii A (2020) Problems of Russian heat supply and ways of their solution. In: E3S web of conferences. mathematical models and methods of the analysis and optimal synthesis of the developing pipeline and hydraulic systems, pp 02003. https://doi.org/10.1051/e3sconf/202021902003
3. Evseev EG (2022) Resource-balanced management of the functioning and development of heat supply organizations. Dissertation for the degree of Doctor of Sciences in the direction 08.00.05 Economics and management of the national economy. M. 2022, pp 388
4. Terentyeva AS (2020) Analysis of the main problems of district heating in Russia at the present stage. Scientific papers: Institute of National Economic Forecasting of the Russian Academy of Sciences, No 18, pp 253–273. https://doi.org/10.47711/2076-318-2020-253-273
5. Verstina N, Evseev E, Tsuverkalova O (2021) Strategic planning of construction and reconstruction of the facilities of the heat supply systems with the use of scenario approach. In: E3S web of conferences. 24th international scientific conference construction the formation of living environment, FORM 2021
6. Novak AV (2023) Report on the results of the work of the fuel and energy complex in 2018 and tasks for 2019 at a meeting of the Government of the Russian Federation [Electronic resource]. https://minenergo.gov.ru/node/14548. Accessed 13 Apr 2023
7. Gavrilenko IG, Khakimov TM (2019) Public-private partnership as a tool to increase the investment attractiveness of housing and communal services. USNTU Bull Sci Educ Econ Econ Ser 3(30):116–127
8. Štreimikienė D, Strielkowski W, Lisin E, Kurdiukova G (2020) Pathways for sustainable development of urban heat supply systems. In: E3S web of conferences, pp 04001. https://doi.org/10.1051/e3sconf/202020804001
9. Report on the state of heat power and district heating in the Russian Federation in 2020. Information and analytical report of the Ministry of Energy of the Russian Federation. 2022, pp 98
10. Federal Law "On Heat Supply" dated 27.07.2010 N 190-FZ (latest edition). [electronic resource]. https://www.consultant.ru/document/cons_doc_LAW_102975. Accessed 12 Apr

2023
11. Bortalevich SI, Loginov EL et al (2018) Problems of forecasting critical technical situations in the UES of Russia taking into account smart grid. In: Problems of safety and emergency situations, No 1, pp 30–37
12. Grabchak EP, Loginov EL, Meshcheryakov SV, Chinaliev VU (2020) Approaches to the integration of information on resource and financial flows in the fuel and energy complex in the conditions of digital transformation of management systems. Management 2:13–19. https://doi.org/10.26425/2309-3633-2020-2-13-19
13. Verstina N, Evseev E, Tsuverkalova O, Kulachinskaya A (2022) The technical state of engineering systems as an important factor of heat supply organizations management in modern conditions. Energies. T. 15, №3. https://doi.org/10.3390/en15031015
14. Zorin SV (2012) Improving the efficiency of heat and energy resources management. In Bull Izhevsk State Agric Acad 4(33):91–94
15. Mendelevich VA (2014) Introduction of modern automatic control systems is one of the best ways to increase the efficiency of operation of thermal power equipment. Autom IT Power Eng 12(65):16–21
16. Rudnitsky G (2023) Digitalization of heat supply: reduction of losses and reduction of accidents. IT Manag Mag [Electronic resource]. https://www.it-world.ru/cionews/security/188651.html. Accessed 14 Apr 2023
17. Decree of the Government of the Russian Federation No. 1632-r dated 28.07.2017. On the approval of the program "Digital Economy of the Russian Federation" (2017)
18. Decree of the President of the Russian Federation dated 09.05.2017 No. 203 "On the Strategy for the Development of the Information Society in the Russian Federation for 2017–2030"
19. Solov'ev VI (2019) Digital transformation of municipal heat supply systems. Inf Math Technol Sci Manag 2(14):52–61. https://doi.org/10.25729/2413-0133-2019-2-05
20. Astratova GV, Rutkauskas TK, Rutkauskas KV, Klimuk VV (2021) Creating an environmentally safe and reliable heat supply system through the introduction of energy-saving technologies in housing and common utilities services. In: E3S web of conferences, pp 04020. https://doi.org/10.1051/e3sconf/202126504020
21. Goncharov E (2023) Heating networks in 2020: advantages of digitalization. Digital Economy. [Electronic resource]. https://www.comnews.ru/digital-economy/content/208954/2020-09-07/2020-w37/teploseti-2020-preimuschestva-cifrovizacii. Accessed 14 Apr 2023
22. Guryanov LV, Ugrevatov AY, Kulikov AN (2022) Digitalization of small thermal power engineering in modern conditions. Myth or reality? In: Industrial and heating boilers and Mini-CHP, No 4 (73), pp 18–23
23. Bank B (2023) What is digitalization?. Vlast.kz. [Electronic resource]. https://vlast.kz/corporation/24539-cifrovizacia-biznesa.html. Accessed 15 Apr 2023
24. Garifullin BM, Zyabrikov VV (2018) Digital transformation of business: models and algorithms. Creat Econ 9:1345–1358. https://doi.org/10.18334/ce.12.9.39332
25. Ovchinnikova OP, Kharlamov MM, Kokuitseva TV (2020) Methodological approaches to improving the efficiency of managing digital transformation processes at industrial enterprises. Creat Econ 14(7):1279–1290. https://doi.org/10.18334/ce.14.7.110615
26. Zaichenko IM, Smirnova AM, Borremans AD (2018) Digital transformation of industrial enterprise management: the use of unmanned aerial vehicles. Sci Bull South Inst Manag 4:76–81. https://doi.org/10.31775/2305-3100-2018-4-76-81
27. Vyugina DM (2016) Digital media business strategies in the context of changing media consumption. Mediascope, No 4, p 20
28. Methodological recommendations on the digital transformation of state corporations and companies with state participation. Approved at the meeting of the Presidium of the Government Commission on Digital Development, the Use of Information Technologies to improve the quality of life and business conditions on November 6, 2020. Ministry of Digital Development, Communications and Mass Communications of the Russian Federation, Moscow, 2022, pp 216 (2022).

29. Yarmiev R, Chernikova A (2023) Digitalization of heat supply as planning of its development. [Electronic resource]. https://nsportal.ru/ap/library/nauchno-tekhnicheskoe-tvorchestvo/2023/02/01/tsifrovizatsiya-teplosnabzheniya-kak. Accessed 11 Mar 2023
30. Departmental project "Digital Energy". [Electronic resource]. https://minenergo.gov.ru/node/14559. Accessed 01 Apr 2023
31. Materials from the website of the Ministry of Energy of Russia. Departmental project "Digital Energy". [Electronic resource]. https://minenergo.gov.ru/node/14559. Accessed 16 Apr 2023
32. Materials from the information portal RosTeplo.ru. [Electronic resource]. https://www.rosteplo.ru/nt. Accessed 27 Oct 2022
33. Materials from the official website of JSC Rusatom Infrastructure Solutions. [Electronic resource]. https://rosatom-teplo.ru/news/1044. Accessed 18 Apr 2023
34. "Digital heat" of Russian origin (2022) JUICE. 2022. [Electronic resource]. Access mode: "Digital heat" of Russian origin. News: 23 May 2022. Accessed 14 Apr 2023
35. Balakhonova IV (2021) Assessment of digital maturity as the first step of digital transformation of industrial enterprise processes: monograph. PSU Publishing House, Penza, pp 276. ISBN 978-5-907456-72-3

Railway Transport Digitalization: Development Methodology and Effects of Digital Implementation Processes

Natalia A. Zhuravleva⑩ and Tomas Kliestik⑩

Abstract The most important priority in the development of transport worldwide is the creation of high-tech, smart and assessment services for the passengers and goods transportation. Digital transformation of transport systems is a response to a request by the passenger and cargo owner for obtaining a new "value" of transportation. This "value" changes the essence of a traditional transport service to a service that provides for high mobility of the population and comprehensive freight chains. Speed, accessibility, safety and cost have become the primary value of transportation services. In Russia, rail transport, as the safest, most environmentally friendly, and useful, is the basis for the formation of the new value of the transport service. Purpose of the current research is to develop a methodology for digitalization of the customer-oriented services forming processes for the of passengers and goods transportation based on railway transport and to evaluate its' effectiveness. Research methodology is based on the concept of market value; on the concept of "competition in the digital economy"; on the definition of "Big Data" as the basis for the new value propositions; on the concept of the value network (Valuenet works), as a set of interconnected innovative processes' chains and technologies, united in a network of services and assets. The results of the study are justification of the basic methodological provisions of the railway transport digitalization: development of platform solutions for the transport service production corresponding to the "value" of the client; competition as a network of partnerships without changing ownership of assets; Big Data as a digital asset in creating new offers and customer services; digital innovation accelerating the commercialization of new ideas and the market entry of new modes of transport and services. A sequence of digital solutions implementation for the transition of railway companies to digital business models has been formed. Such business models focused on increase in the mass of incomes through new value propositions for the client and increase in the number of clients. Models for assessing the effects of the railway transport digitalization are systematized. Its'

N. A. Zhuravleva (✉)
Emperor Alexander I St Petersburg State Transport University, 190031 St Petersburg, Russia
e-mail: epro@pgups.ru

T. Kliestik
University of Zilina, 010 26 Zilina, Slovak Republic

applicability to digital technologies at different levels of business processes of a railway company has been proved. Recommendations on the use of a methodology for assessing the effects of digital technologies implementation for interaction of supply chain participants based on agent-based modeling are given.

Keywords Value of transport service · Platform solutions · Mobility · Multimodal transportation

1 Introduction

The Russian transport system is represented by five modes of transport (road, air, sea (water), railway and pipeline), greatly differing in terms of it's infrastructure development level, provision of modern vehicles and technical equipment of the transportation process in general. This is due to a number of circumstances: geographical, spatial, historical, regulatory and managerial, which complicates delivering effective population mobility systems and supply chains. At the same time, different regions of the Russian Federation have completely different transport accessibility, and the recent geopolitical changes have significantly changed the directions of transportation, which led to under loading of some directions and overloading of others. Practice and the current trend in the infrastructure development of each mode of transport for individual programs and projects increases competition only within one or two modes of transport, does not improve the conditions for transportation in general. However, on the contrary, it leads to an increase in the cost of transport services. This forces transport organizations to look for the ways out of this situation, using the possibilities of organizing mixed (multimodal, seamless) transportations, introducing innovative modes of transport and management systems for it. However, in order to ensure high mobility of the population and integration of cargo flow chains that can reduce time of transportation and reduce its cost in modern conditions, a completely different conceptual approach is needed to solve the problem of increasing the efficiency of the Russian transport system. First, it concerns railways, capable of transporting goods and passengers at any distance, just in time, with a high level of safety and at an affordable price. It explains the relevance of the current research.

Transport system development of the country, region or international transport corridors has always been taken into account when substantiating programs and projects for the economic growth of any economy. Today, this process is undergoing significant changes, which relate not only to technical and technological solutions, but also to a significant change in the business models of transport organizations. The transition to a "digital" economy is an objective business need for effective changes that can increase its competitiveness and ensure "winning in the long run", i.e. sustainable development of both their own business and country's economy. The meaning of organizations business models digitalization is to shift the priorities of their activities. Client and production of a product or service that meets his "value" needs–becomes the main priority. Competition turns into a network of partnerships without changing asset ownership relationships. Big Data allows you to analyze all information about consumers, to exclude intermediaries in sales. Innovation enables organizations to commercialize new ideas and quickly launch production, changing traditional sales markets.

The logic of transport systems digitalization meets the requirements of its sustainable development (ESG concept), i.e. growth of environmental friendliness–reduce the burden on the environment, accelerate the socialization of transport services, primarily accessibility and speed, reduce the cost of transportation, as well as the corporate responsibility of the transport business.

The regulatory framework of this research is represented by the Russian legislation regulating the digitalization of the Russian economy, as well by the international legislation regulating the interstate transportation along international transport corridors. The information basis of the research is the data of the Federal State Statistics Service, of the Government of the Russian Federation, of the Ministry of Transport of the Russian Federation, of Russian Railways, of Russian and foreign transport and logistics companies, information resources of RosBusinessConsulting and INFOLine.

Analysis of the status and development prospects of digital processes of Russian railways is based on open research by developers and specialists of Russian Railways, research and design transport institutes, and the authors' own developments.

The following data of logistics indices have been used: efficiency of customs and border clearance; quality of trade and transport infrastructure; ease of organizing supplies at competitive prices; frequency of shipments reaching recipients at scheduled or expected delivery times [1–3].

2 Methods and Data

2.1 Basic Conceptual Provisions Adopted in the Research

In this research, digitalization process of railway transport is built in accordance with the concept of market value, in which the economy is managed by the customer [4–7].

Customer forces manufacturer of goods and services to produce a product according to their requirements [8].

Manufacturers are forced to look for opportunities in order to gain competitive advantages not within their own business, but in cooperation with other businesses, offering the customer the required value of the product [9].

At the same time, the concept of "value" is not unequivocal. Different authors use such concepts as value, price, worth, asset, treasure, wealth, etc. in determining the value of a market offer, without disclosing the content of these terms, which makes it problematic to consider their definitions identical [9].

Customer perceives value of goods or service in different ways, evaluating the benefits of ownership, use, perception. However, the common feature is the ratio perceived by the customer between all the benefits and costs (including non-financial) associated with the acquisition and use of the proposed goods (service). In the current research, we consider a passenger, cargo owner, shipper (forwarder and transport agent), state (budgets of different levels) and personnel of transport organizations as a "client for whom a new value proposition is being formed".

A transport service (services for the goods and passengers transportation) in modern society and economy acquires "value" just when it provides the proper level of safety, mobility for the passenger, optimal for the cargo owner time and cost of cargo delivery, and a positive impact on the growth of the country's GDP. The transport systems of the country, regions, cities and transport corridors can provide this "value" only if there are "seamless" transportation technologies, i.e. door-to-door transportation by various modes of transport with a minimum transfer (transshipment) time, optimal cost for all customers, according to a single document and the required set of related services.

This corresponds to the concept of a value network (Valuenet works), described in 2002 by Verna Allee as "complex sets of technical and social resources that interact with each other and create economic values" [10], as well as by Cinzia Parolini in the book "The Value Net: A Tool for Competitive Strategy" [11]. The term Valuenet itself suggests the use of set of related operations that generate new or significantly improved value for the customer. The value proposition focuses both on the client and on the values or set of elements that make up [12]. At the same time, each element is considered separately, which allows you to separately assess the risks and add new elements with the help of innovations. The value network can be viewed as a set of interconnected chains of processes and technologies that are combined into a network of services and assets. Each service and resource is a node on the network. When using different combinations of nodes, it is possible to provide the client with different options for travel routes or delivery of goods.

As for transport, this set of related iterations can be ensured by synchronizing the introduction of digital technologies into the processes of goods and passengers transportation by different modes of transport. While the new "value proposition" can be both multimodal (seamless) or within one mode of transport, for example, road transport (transportation of a freight container by road over a distance of more than 600 km and delivery by road of the "last mile"). A very important set of the value networks is generated by transport organizations–owners of transport infrastructure, such as Russian Railways–the owner of railway infrastructure and traction, AVTODOR, seaports, airports. It is on the basis of their infrastructure that value is created and delivered with the help of a value proposition. Each transport organization–a provider of transport services is in a state of specific and interspecific competition in the transport market, which depends on the changing commodity market, geopolitics and geo-economics. At the same time, it is impossible to change the transportation infrastructure as quickly as the needs in the directions and structure of transportation change is. That is why digital technologies are so important when designing, building and launching a new transport infrastructure.

Since 2022, a number of goods transportation markets that fell under the sanctions, have significantly narrowed. While in other types and directions of transportation has appeared and is growing a shortage of carrying capacity and capacity of infrastructure and vehicles. Consequently, transport companies are changing their operating models, and transport infrastructure companies are changing business models. A significant sign of the transition from analogue business models of transport organizations focused on saving transportation costs, which ensured stable profit growth, is the transition to digital business models focused on growing the mass of profits through new value propositions for the client and a corresponding increase in the number of customers.

The second conceptual provision of this research is the change in the concept of competition in the digital economy [13]. In the analog era, there were only two options: competition with companies that produce goods or services with the same consumer properties, and association within the framework of industrial cooperation. The digital economy makes possible asymmetric competition or digital "disintermediation" (the process of eliminating intermediaries from transactions between buyer and seller). This enables the transition of the organization to a platform business model, which, as a technological basis for the provision of services/content, connects service providers and end users.

Platforms consist of many levels and trajectories, forming different sets of operations that form a service, which allows customers to choose the optimal service for themselves and makes it difficult to exclude certain players, that is, competitors. By generating a network effect, platforms provide companies with the most important competitive advantages: increasing the value of a service for a user (expanding the set of offers, choosing the optimal service, personalizing services, etc.). At the same time, the company increases the influx of customers due to direct or indirect network effects [14, 15]. Direct network effects can be clearly observed on online platforms, where the value for each user already represented on it increases as additional users join the platform [16]. Indirect network effects are manifested in the provision of multifunctional services, when all participants in the platform offer receive a positive effect from the provision of one type of service [17]. Thus, competition in the digital economy turns into a competition of sets of values. This is a very important conclusion for understanding the processes of digital transformation of transport organizations in the transport services market.

First, digital competition can be found in mixed (multimodal) transportation, built on competitive cooperation. For example, rail transport is more efficient for longer distances and trucking is focusing its business on the "last mile", which provides door-to-door transportation.

Competitors and partners in various transport projects change their roles: the cargo can be transported by the consignor's operators or by the consignee's operators, or by a transport agent.

Key assets can be used as an external network of partnerships without changing the ownership of assets, which allows railway transport, whose property is limited in circulation, to use its logistics infrastructure in multimodal transport processes. Thus, a freight operator company that just provided services for the provision of wagons (containers) previously can purchase locomotives and compete with Russian Railways as a carrier (now on non-public tracks). Another example is the marketplace developed by Russian Railways–RZD Market, with a delivery by railway, which, as a digital service, optimizes the solution for finding, selling and buying goods with delivery to the end consumer. In the same way, the problem of increasing the mobility of passenger traffic is solved. Digital passenger railway platforms integrate not only different modes of transport, but also services for the passenger: food, entertainment on the road, transfer, hotels, etc., thus facilitating the exchange of value for all service providers. In this case, it becomes necessary to cooperate with a company that is at the same time a direct competitor, for example, long-distance passenger transportation can be carried out by both JSC Federal Passenger Company and JSC TC Grand Service Express on the same route St. Petersburg–Moscow. At the same time, companies need to fit into the transportation schedule and agree on a pricing policy.

An analysis of programs and projects for the digitalization of intermodal railway transportation shows that in the next 2–3 years, large forwarders, agents and supply chain operators will compete not as separate organizations, but as regional or corporate digital ecosystems. By 2024, supply chains and their platforms will be transformed/redesigned based on the principle of modularity, using innovative digital business models. By 2026, more than half of supply chain organizations will use artificial intelligence (AI) and machine learning (ML) to empower decision-making.

The following conceptual position is based on the fact that if the digital economy is formed by Big Data, then Big Data provides the formation of new transport services values: gathering, storage, processing, dissemination and analysis of data [14, 18, 19]. Data becoming the most important asset of organizations and differs from other types of assets because it is conditionally "out of competition"[20], ensures economies of scale are reached and generate direct and indirect network effects to developers and users [21].

Big Data generates an environment where market participants (primarily platforms) compete "for" the market rather than "within" the market. The unique nature of data, network effects, economies of scale all combine to dramatically increase the advantage of companies actively pursuing digital transformation. Data gathered and analyzed by organizations in one market can give a competitive advantage in other markets and help to generate synergies. Using data to improve forecasts increases the value of offers for users and helps to attract additional users. This series of cause and effect events, i.e. positive feedback loop, referred to as data network effects [22].

Basically the non-competitiveness of data makes it possible to use it several times without depleting and in any organization simultaneously [23, 24]. In many organizations accumulation of large amounts of data on related markets and products, on companies–"disruptors", is a basic component that helps to change their business models and enter new markets or areas of activity, thus achieving fixed cost economies due to business scale. In this case, the concept of competition also changes: when organizations begin to participate in vertical integration, when two or more successive stages of manufacturing and/or sale of services are combined under one management [25].

All activities of transport organizations based on movements of goods and passengers information in real time ("data in motion"). After the appearance of tasks associated with a particular data type management of such data each time moved to a new stage. The need to analyze relationships between the performance of all processes in the company stimulated emergence and development of relational databases. At the time when the value of the transport services has shifted towards the growth of mobility, "seamlessness" and optimal cost, i.e. customer focus, the need for unstructured data and its processing based on natural language appeared. By transporting, companies generate data everywhere and constantly, turning it into a critical asset. In this context look different digital platforms on which business process management, version control, information recognition, text management and collaboration are implemented. One of such platforms in the Russian Federation is the State Information System of Electronic Shipping Documents, which provides the consignor, carrier and consignee with an electronic consignment note, an accompanying list, an order, a waybill, an application and an electronic charter agreement. Today, organizations are less likely to talk about the high cost of digital platforms–Internet technologies, visualization and cloud computing have significantly reduced the cost of computing cycles and data storage [14]

Transportation data integrated into vertical and horizontal information arrays, which increases the importance of value network externalities, affecting the efficiency of the companies themselves. An example of such integration is digital platform of the Russian logistics company Eurasian Agrlogistics, which is a freight forwarder by its legal status. The platform provides organization, transportation, including customs clearance, border crossings and other services of the agro-industrial complex from / to the Russian Federation through Iran to India, involving in circulation goods of Central Asian countries (Uzbekistan, Tajikistan, Kyrgyzstan, Turkmenistan and the Persian Gulf countries). The digital network of this project is built on a flexibly scalable cross-platform IT solution with client services: personal accounts with electronic certification, customs clearance, order management, shipment tracking capabilities, an electronic exchange for large and small wholesale, integrated with foreign IT systems of countries exporting Eurasian goods. Internal services provided by a system of end-to-end supply chain management with the prospect of reaching the 4/5PL level, which provides for electronic purchases/sales (network business module). On this platform, network effects and economies of scale complement each other to help ensure the competitive advantage of providing agricultural goods to the populations of several countries and regions. Network effects mean that each additional user brings more benefit or value to all users, while economies of scale mean that the average cost of serving users goes down. Thus, the benefits to consumers increase even as costs decrease. From a technical point of view, at this level of fixed costs, there are no economic restrictions on the expansion of digital objects (attracting a new number of customers from different countries on organized routes).

Digitalization process of any activity cannot be fully described without considering the conceptual provisions for the development and implementation of innovative technologies and solutions. Innovation changes the cost of a value proposition, while digitalization of the innovative proposals developing processes greatly simplifies the emergence of a new product or service. Innovations are not only "big bangs": they include all novelties that have value or add value to an already existing offer [26]. In the digital economy innovations based on constant experimentation and acquisition of new knowledge, which distinguishes them from traditional innovations that must be turned into a new finished product that is successfully sold on the market.

The concept of digital innovation is based on the principle of continuous experimentation of not solving but eliminating a problem. This is the central thesis: company should focus on customers and needs. Concentration on the problem allows creating several solutions. At the same time, decision-making depends on the results of testing, carried out quickly and without much cost. For example, by creating digital twins, computer simulation or virtual reality technology. The concept of digital innovation, in addition to various methods for developing experiments, offers several ways to scale innovations [27]. You can scale the created minimum viable prototype (MVP–Minimum Viable Prototype). It is developed with a specific set of capabilities to get useful feedback from customers [28]. Scale can be achieved by launching an MVP everywhere, when a company needs to take advantage of network effects. For example, a platform business model for organizing freight transportation in the Belarus-Russia-Kazakhstan transport corridor. Quite often happens when innovation can be launched only as a complete product, for example, a self-driving car. At the same time, controlled product output in selected markets or segments is still being tested.

One of the innovative transport systems types are intelligent transport systems (all types of transport, including the urban public transport network). The most popular of them is a self-driving car. However, innovative processes being introduced more widely in the development of digital twins of vehicles and design solutions for the construction of transport infrastructure. Experimenting with it makes it possible to reduce the costs of designing, maintaining and operating transport infrastructure facilities of all types of transport. Very important innovations in "digital" predictive repair of vehicles and transport infrastructure facilities, in creation of digital (intelligent) terminals (passenger, cargo), checkpoints across the borders of international transport corridors, in introduction of machine learning technologies, drones, surveillance and predictive response systems to ensure transport security and many other processes that significantly affect the value of transport services.

Innovations in transport, first, should ensure the intellectualization of the multimodal (combined) transportation mobility, which forms a new value of the transport service for passenger and cargo owner. In terms of passenger traffic, it forms transport accessibility, a single ticket and the services required during the trip. In terms of cargo, transportation provides competitive speed of transportation, optimal cost for business and proper supply chain services.

Implementing innovative technologies can destroy traditional supply chains: changing the nature of existing contacts, role and value of data, level of value creation at each segment, destroying old and creating new operational and business models of transport and logistics companies. The degree of destruction depends on the scale of the innovation's introduction–digitization of individual technologies and processes or of business process in general. Digital base of innovation increases the difference between traditional supply chains that have been digitally enhanced and truly integrated, reinvented, those ones which DNA is basically digital.

3 Research Results

3.1 Analysis of the Compliance of Railways' Digitalization Processes in Russia with the World Level of Transport Services Quality

In Russia, as in many other countries, value of a transport service is determined by the level of its transport system development: transport accessibility, quality and capacity of infrastructure, coverage of the territory by transport networks, quality of transportation services, etc. In this case, digitalization of railways becomes an essential element of the digital economy, moving the value added hubs of transport services into the digital resource forming industry and end-to-end digital processes based on railway transportation. There is a number of reasons for this: in particular, the maximum coverage of the territory of the Russian Federation (operational length 85.6 thousand km; share in freight turnover–46% (including pipeline transport), in passenger turnover–26.4%.), high efficiency (single index of railway tariffs–103.7; in general for freight transportation–104.8), environmental friendliness of vehicles (85% of transportation on electric traction), ability to transport goods of all types and sizes, independence from the weather conditions etc. All of these reasons well documented in the literature and statistically confirmed [29]. At the same time, Russian railways are not only a transport network on the territory of Russia, but the infrastructure of several international transport corridors: the Eurasian North–South and Trans-Siberian corridors, accesses to the Northern Sea Route, pan-European corridors No. 1, No. 2 (Pan European Transportation corridor, PETC) #2 (Berlin-Warsaw-Minsk-Moscow-Nizhny Novgorod), No. 9#9 (PETC 9) (Helsinki-Moscow-Black Sea-South Europe). The railway connects accesses to the northeastern provinces of China through Russian seaports in the Primorsky Territory with the ports of the Asia–Pacific region–Primorye-1 and Primorye-2.

Since 2021, the new geo-economics has significantly changed the supply chains of goods formed both in the Russian Federation and in transit cargo passing through its territory. Seriously increases competition in international corridors bypassing the transport routes of Russia. Thus, TRACECA, Eastern Europe (PETCs 4, 7, 8 and 9)–the Black Sea-Caucasus-Central Asia and the South corridor (South-Eastern Europe (PETC 4)–Turkey-Iran with two branches to Central Asia and China [30].

Client chooses routes provided with coordinated legal regulation, with a unified tariff policy and a single customs space. At the same time, in the current circumstances, competitiveness is not always determined by low transportation tariffs. Time is of the utmost importance as an effectiveness indicator of the operating models of cargo owner and carrier.

By these and other reasons, led to the digital transformation of Russian railways, the strategic direction of their development has shifted towards the formation of a competitive transport service that meets international standards of transportation quality. Table 1 shows the results of the analysis of the digitalization processes of Russian railways compliance with the level of digital processes of the railway companies in the leading railway countries (USA, Germany, China).

Analysis of digital projects of the largest railway companies and Russian railways shows the full ability of the latter to provide the proper quality and value of the transport service to customers. At the same time, the problem for all railway companies remains their ability to maintain the required level of profitability of transport services in the face of growing specific, interspecific and spatial competition in the transportation markets. For Russian railways, a competitive advantage is the creation of platform solutions that become a technological, informational and innovative basis for the provision of freight and passenger transportation services, unite service providers (transport and logistics companies of all types of transport, freight car operators, service providers) and end users (passengers and cargo owners).

3.2 Digital Platforms of Russian Railways and Its Role in Digital Competition in the Transport Services Market

Digitalization of the transport industry in the Russian Federation is carried out in accordance with the National Program and the national project "Digital Economy of the Russian Federation" [33], with the Transport Strategy of the Russian Federation until 2030 with a forecast for the period up to 2035 [34]. The digital infrastructure of JSC "Russian Railways", as a state-owned company, complies with the goals and parameters of the Transport Complex Digital Platform of the Russian Federation, the digital transformation program, methodological recommendations for the digital transformation of state corporations and companies with state participation, and other documents regulating national industry digitalization programs [35–37]. The logic of digitalization of railways is aimed at creating an economic device in which new value propositions are formed for the client, digital data becomes a key asset, and digital innovations ensure the growth of mobility and optimization of transportation costs, which increases the country's competitiveness, the quality of life of citizens, and ensures economic growth, and national sovereignty.

Digital transformation of Russian railways began with fundamental changes in logistics and the main areas of transport goods manufacturing (services for the transportation of goods and passengers), forming fundamentally new methods of communication, connections with customers and counterparties, and new methods of creating added value [38].

Table 1 Digital technologies implemented in the processes of transportation in Russia and the leading railway companies in the world

Processes	Best practices for railway companies	JSC "Russian Railways" (digital platforms implemented by the company)
Focus on customer		
Sales forecasting	Statistical analysis and machine learning to pyritize the customer base and reduce customer churn	Customer churn management within the customer experience management platform
Optimization	Predictive models for demand forecasting and price optimization using price elasticity for different customer segments	Dynamic pricing within the customer experience management platform
Use of the rail network	Using simulation to improve network usage efficiency under different infrastructure constraints	Information and control module for the development of normative, daily traffic schedules and windows on the infrastructure
Sales management	CRM (customer relationship management) system for automating customer interaction strategies	Integrated system of interaction with clients (freight transportation): Client's personal account; Electronic trading platform "Freight transportation"; Calculation of the cost and environmental friendliness of freight transportation
User experience	Personalized communication with clients, including using real-time data Ability to plan a trip and buy tickets for several modes of transport	Customer Relations Management System Multimodal transportation system (railway, electric trains, buses, air)
Big data		
Rolling stock maintenance	Using predictive analytics to predict the need for repair/replacement of rolling stock components in order to optimize maintenance costs	Predictive analytics of technical condition based on mobile diagnostics data Predictive analysis of the technical condition of freight cars
	Autonomous sensors for monitoring the technical condition of rolling stock units and rail infrastructure	The use of mobile diagnostic tools and the use of predictive analytics of the technical condition of the composition based on the data received Implementation of stationary systems for complex diagnostics of railroad switches

(continued)

Table 1 (continued)

Processes	Best practices for railway companies	JSC "Russian Railways" (digital platforms implemented by the company)
Fuel efficiency	Selection of the optimal mode of train movement in order to save fuel based on complex multifactorial models simulating various conditions on the way (type of locomotive, train length, etc.)	Information and control module for managing traction resources
Asset management	Digital control for the whole train movement (departure/loading, sorting, arrival/unloading) Analytics and data visualization providing real-time insight into the health and reliability of assets (OEE, causes of loss, etc.)	Monitoring of infrastructure state predictive diagnostics by the objects Information system for managing the life cycle of track machines and mechanisms
Planning	Automated systems for planning operational indicators (including calculation of the required fleet of locomotives and its availability), schedules and route network, sales	Information system for planning the work of locomotives Digital model and predictive analytics (forecast) on the technical condition of the tracks
Innovation		
Autonomous control mode	Autonomous control mode of the locomotive Autonomous shunting locomotives-robots Fully autonomous trains without a driver for the delivery of goods over long distances	Implementation of the "Automatic train driver" system on hump locomotives Driver assistance system for driving (technical vision) Ensuring the uncoupling of wagons during the dissolution and fully automatic dissolution of wagons on marshalling yards
Application of robots for routine tasks	Automation of the process of generating operational reports for making prompt decisions and responding to emergency situations	The use of software robots for auxiliary operations, for user support systems, and information and reference systems Automation of route preparation based on the current train schedule (including in emergency situations) Automation of dispatch control Automation of processes at the station

Compiled according to: [31, 32]

In the Russian market of transport services, digitalization processes are most active in JSC "Russian Railways" as the owner of the railway infrastructure and the largest freight and passenger operator in the country. The company combines two types of transformation–manufacturing and business model. This transformation has a clear economic goal: the growth of the rate or mass of profit in the face of changing commodity markets, changes in the volume and range of goods presented for transportation, and expectations of high mobility of the population. The goal defines the business landscape for the digital transformation of railways, creating a new "value" of transportation services, changing the most important areas of activity of transport and logistics companies: customers, competition, data, and innovation.

At the same time, competitive environment in the transport market is active both in the railway transportation segment (more than 10 thousand railway operators) and in relation to other modes of transport. The program "UAVs for Passengers and Cargo" is implemented on vehicles, digital processes for managing traffic flows (the program "Digital Management of the Transport System of the Russian Federation", the program "Seamless Freight Logistics", which is already reducing the time for customs procedures by several times, digital ticketing systems are being introduced in many agglomerations systems that reduce the time of waiting and passing through control procedures on all modes of transport (the program "Green Digital Passenger Corridor")), etc. [39, 40].

Digital transformation goal for the "Russian Railways" is to create conditions for the realization of the "desire" of customers to transfer most of the transportation (trips) to railway transport. However, this is possible only if the client is sure that such a service reduces time and optimizes cost. To win the competition, one must either team up with a competitor and offer a mixed service, or transport cheaper, faster and with better service.

Competitive value propositions for railway transportation based on the service model of the Russian Railways holding, shown in Fig. 1.

The service model has been "digitized" on eight digital platforms–complexes of interconnected technological solutions for the interaction of participants in the transport market (Table 2).

On the customer platforms multimodal freight and passenger transportation, digital services based on the modular principle. It ensures good composability, in which the resources and processes of the supply chain (transportation) divided into "building blocks" ready for reuse and reconfiguration. Supply chain (trips) modularity helps to standardize and develop interoperable platform solutions and services that can be widely used by new transportation participants as "building blocks" of their digital infrastructure (mobile applications, gaming and information services, face recognition system, etc.)

By creating these standardized "building blocks" in the form of generic modules and services, rapid scalability and connection of new participants to the supply chain ecosystem is ensured. At the same time, it should be taken into account that digitalization changes the solutions to the "classic" tasks of the railway, such as expanding the network, improving safety, increasing freight and passenger traffic, attracting

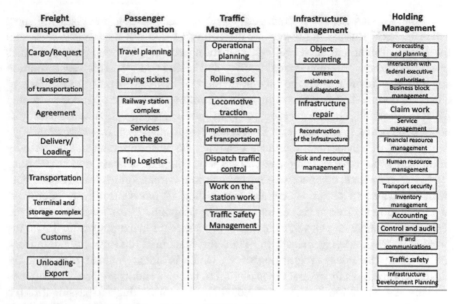

Fig. 1 Service model of Russian Railways Holding [38]

Table 2 Digital transformation of JSC "Russian Railways"

JSC "Russian Railways" services		Subsidiary services	Services for other market players		
Internal customers	External customers	Subsidiaries and affiliates	Government	Partners	Counterparties
Functional customers Employees	Passengers Consignors	Subsidiaries and affiliates	Ministries and departments	Scientific institutions Funds/ Startups Technological partners	Counterparties Customers
Digital platforms					
Digital corporate culture	Multimodal passenger transportation	Line infrastructure operator	Multimodal freight transportation	Transportation process management	Process and Regulatory Optimization
	Traction rolling stock	Non-manufacturing processes	Logistics e-commerce operator	Transport and logistics hubs	

new customers, adding a range of new digital services (multimodal, technological, managerial), adaptable to market conditions and needs passengers and cargo owners), which positively affects the financial performance of all participants in the transportation.

3.3   Below Are Cases of Platform Solutions of JSC "Russian Railways"

Customer-orientation passenger transportation. The Federal Passenger Company (FPC) uses digital technologies, in particular, the Internet of Things, Big Data, to analyze passenger flows and predict the need for transportation on existing and future routes. These technologies provide data options for the development of multimodal routes, CRM (Customer Relationship Management) implementation systems and customer data management, forming offers of various sets of services for each route. The main services on the Multimodal Passenger Transportation platform can be received by a passenger in the form of a comprehensive door-to-door transportation with the involvement of third-party carriers and on the electronic platform of additional services (class of service, media content during the trip, insurance, etc.). Growth of commission income and the consolidation of traffic in one place for the future monetization of all services and growth of passenger loyalty are the bonus for the company.

Customer-orientation of freight transportation implemented on the multifunctional platforms "Multimodal freight transportation", "Transport and logistics hubs" and "Logistics e-commerce operator". Platforms generate an integrated customer interaction system–CRM; form a distributed registry of transportation, allow to conclude smart contracts. "The Freight Transportation electronic trading platform gives shippers the opportunity to order transportation in rolling stock of any type from anywhere where there is Internet access, and pay for it from a single personal account or bank payment [41].

The platform processes all information about the client and history of his interaction with JSC "Russian Railways" in order to prepare personalized offers for him, including global services. Sale of complex transport services is calculated, applications confirmed and inquiries received on a convenient remote portal. Thus, based on history of relationships and their needs an individual approach to customers implemented. This significantly affects the growth in the share of rail transportation in the total freight turnover and income from additional services.

The locomotive complex (Traction rolling stock) is the most important digital platform for the company, which provides a solution to the problems of increasing transportation costs. The platform implements solutions for the already introduced innovative technologies of the Smart Locomotive and Trusted Environment projects, within the framework of which the issues of automatic diagnostics of the technical condition and creation of digital "twins" are solved. It provides predictive diagnostics of the state of the locomotive, planning and control of the locomotive complex. These platform solutions are associated with the need to improve the reliability and safety of rolling stock and significantly reduce operating costs. Accurate understanding of the current state of locomotives due to remote and automated diagnostics, carrying out repair work not according to the standard schedule, but on the fact of equipment wear and tear, can significantly reduce the cost of servicing rolling stock by increasing the overhaul period and reducing unscheduled downtime.

Due to predictive diagnostics ("Trusted Environment"), the readiness indicators of the rolling stock increase, which makes it possible to rationally use depot resources, reduce the number of electric trains in reserve, and ensure 100% compliance with the traffic schedule. On this platform, a JSC "Russian Railways" trusting environment is formed with locomotive repair plants, service companies, manufacturers and suppliers of equipment for locomotives, including the validity and efficiency of decision-making in the event of responsibility for the failure of components and assemblies rolling stock ensuring, as well as the simplification and unification of information exchange between the participants of the locomotive complex.

Transportation process management. Intellectualization of the railway transport management system, digitalization of the operation of stations and local operation of railways are the main conditions for reducing the time of transportation or increasing the speed of movement. Since transportation is controlled in real time, reducing the influence of the human factor on the efficiency of the system (for example, station personnel on duty) is the main one in this process. The reduction of transportation time also ensures the speed of its planning (recalculation and updating) taking into account new factors (for example, changes in transportation costs). Ultimately, the productivity of locomotives increases and losses from non-scheduled delays and train stops decrease.

Digitalization of railway transportation is widely integrated with information platforms of other modes of transport. The most indicative result of such integration is the work on electronic data exchange "port–road" and "road–port", which can significantly improve the quality of goods loading planning to the ports, eliminating unproductive losses from wagons set aside from the movement on public tracks and non-purpose occupation infrastructure. The pilot project has been implemented in the system of interaction between the Novorossiysk Commercial Sea Port and the North Caucasus Railway. Based on the data received from the port, according to the demand for cargo in the context of the assortment for the planned period, if these cargoes are available on the railway, a forecast for the arrival of trains with demanded cargo for two days in advance generates automatically. The problem of ensuring the rhythm in the advancement and uniform loading of terminals is under solution. It reduces unproductive losses and increases the volume of unloading at port stations.

3.4   Economic Effects of Railway Digitalization

The economic meaning of digitalization, measurement of the new value of a service or product, effectiveness of the implementation of digital technological solutions involves the systematization of evaluation tools: a set of techniques, methods and models of economic evaluation that ensure the reliability of the results of digital transformation projects [42, 43].

From a variety of economic assessment methods, in order to confirm the effects of digitalization of railway transport, we have chosen the following:

- evaluation of absolute and comparative economic efficiency,
- evaluation based on a balanced scorecard,
- evaluation of digital technologies of interaction between participants in supply chains based on agent-based modeling.

The methodology for determining the absolute and comparative economic efficiency of railway digitalization projects is quite traditional and well adapted in the industry. Its essence is in evaluating the effects of investments in a digital project, i.e. standard indicators of absolute and relative economic efficiency, the payback period of investments in the project, indicators that determine the choice of the best of the alternative solutions for the use of a particular technology. Thus, the introduction of innovations in transport is assessed. For example, the acquisition of new generation track repair machines, which allow increasing labor productivity tenfold; installation of the GLONAS system on the rolling stock, etc. This method is used in cases where the project has one specific owner (investment, implementation, operation, monetization of the effect is carried out in one organization). Other participants in the transport services market are not taken into account. Table 3 shows an example of evaluating the effectiveness of projects for the implementation of artificial intelligence in Russian Railways.

As follows from the Table, the resulting "effects" from the introduction of digital technologies are evaluated not only directly (in the amount of income received), but also indirectly (in improving organizational measures, reducing the time for processing documents, etc.), which ultimately will also positively affect implementation effectiveness.

Methodology for evaluating the effectiveness of railway digitalization processes based on the construction of a balanced scorecard (BS–Balanced Scorecard).

The use of this methodology is due to the fact that a digital solution does not always explicitly and directly lead to an economic effect, assessed by a deterministic set of indicators. A number of digital solutions (digital twins, digital services and mobile applications, digital platforms, Internet of things technologies) do not always have a direct economic effect. For example, digital services and platforms are being introduced to increase customer focus and create additional competitive advantages, virtual reality technologies, digital twins are created to reduce staff errors and train employees. BIM technologies are associated with the need to improve the system of R&D and innovation development.

The BS methodology is a comprehensive project assessment that combines the most important areas of company's activity: client, innovation, social status and economic result–everything that we consider in the digital economy. With its help, we evaluate the indicators of changes in financial results and the market value of the company; as well as magnitude of the influx of additional customers by creating a new value of transport services, identification of business processes that have improved in the process of digitalization (for example, replacing routine operations with robotic processes, predictive diagnostics of equipment operation, etc.), and impact of intangible digital assets on the growth of staff productivity, training, development of employees, etc.

Table 3 Artificial intelligence in railway transport

Solution	Effects
The Concept for the Application of Artificial Intelligence in JSC Russian Railways was approved	Systematic work on the use of AI technologies in railway transport has been built
Intelligent processing of measurement results obtained from commercial inspection technical means	The first stage: the use of STCI[a] at 33 points resulted in a reduction in costs by 60 million rubles per year and release of 122 employees
Natural dialogue system–automatic selection of answers to a user's question	Reduce the processing time for requests from users of information resources by 3–5 times; coverage of processing using artificial intelligence in 2022–processing of 40% of requests without human intervention, or with minimal participation, in 2023-50%
Machine vision (shunting locomotives, unmanned trains "Lastochka")–tested	Expected effects: –up to 20% growth of throughput capacity of stations in the future –transition to a three-minute interval between trains on the Moscow Central Circle (MCC) and an increase in passenger traffic by 200,000 passengers per day –saving operating costs by reducing the staff of locomotive crews and increasing the reliability of the electric train
Intelligent assistant shunting dispatcher (prototype)–tested	Expected effect: 20% reduction in the average downtime of a transit wagon with processing

Compiled according to: [44, 45]
[a]STCI-Software and technical means of commercial inspection

Since digitalization most often is aimed at solving the strategic tasks of the company, this methodology allows to link strategic and operational goals, external and internal risks, evaluate the effectiveness of projects that, in terms of direct effects, only increase costs.

Methodology for assessing the effects of introducing digital technologies for the interaction of participants in supply chains based on agent-based modeling. Agency models make it possible to optimize the interaction of each participant entering the digital environment with other participants with whom operations are carried out that describe the request (demand) and the possibility of satisfying the demand (supply). Based on the application of the methodology, predictive estimates of the consequences of implementing digital solutions are obtained as the results of choosing the best options for supplier–consumer interaction in the supply chain management system through information exchange in the digital environment. The main task of the multi-agent model is to ensure the implementation of the goals and objectives of an agent (for example, a participant in a digital platform). Getting into a multi-agent system, the agent tries to fulfill the goal using the methods available to him, which is also pre-set by the specialist developing the model.

When modeling the supply chain, the goal of the agent is to maximize the loading of vehicles, reduce transportation time, and save transport and logistics costs. All this is reflected in the operating cycle of the cargo owner company and in the manufacturing cycle of the carrier (owner of infrastructure and traction and railway operators). As a result of multi-agent interaction, maximum efficiency is achieved for all participants in transport and logistics relations integrated on a digital platform, including providers of various services.

Despite the frequent use of these methods, it is practically impossible to obtain the most reliable estimate of digital effects using only one method. Digitalization generates complex effects that are difficult to formalize as the very concept of "transport service value". The complex effectiveness of economic, technological, social and environmental effects should be subject to assessment. An example is the performance of the target portfolio of IT projects of the RZD holding:

1. The effectiveness of the company's digital projects is at the level of 0.86 on a scale from 0 to 1.
2. Due to the added IT projects, support for the activities of all service blocks of the Digital Railway model is noticeably increasing.
3. The rationality parameter of the target portfolio of IT projects of the Russian Railways holding corresponds to a value of 0.81 on a scale from 0 to 1.
4. The rationality of the target portfolio of projects has increased due to the addition to the target portfolio of IT projects included in the IT Development Strategy of Russian Railways for the period up to 2020 and IT projects. Its results have a high contribution to the development of service units, and the risks are at low and medium levels [38].

The digitalization of railways is synchronized with the digitalization of the Russian economy in general. Table 4 shows the statistics that reflect the level of this synchronization.

As can be seen from the table, the process of digitalization of Russian railways, having a number of advantages in terms of the growth rate of the digital technologies introduction, effectiveness of new value propositions, intellectualization of processes and labor, is seriously behind national indicators in terms of the number of advanced domestic technologies introduced and investments in their development. This is due to the fact that when the company was reformed in 2003, R&D departments and activities were excluded, so that the transportation tariffs could include expenses exclusively related to it. Now the company mainly acquires new, innovative technologies in the market.

4 Discussion

Methodological approach adopted in the research is based on the conceptual provisions of the digitalization of the economy, presented in the researches of foreign scientists on the theory of market value, in particular, [4, 8]; value creation networks

Table 4 Values of indicators of digital transformation in the Russian economy (in general) and in the Russian Railways holding

Indicators	The value of the indicator in the holding "RZD"	Significance for the Russian economy
Growth rate of R&D costs in the priority area "digital systems and technologies" (2016–2020), times	12,8	0,98
Volume of R&D performed in the priority area "digital systems and technologies" per 1 employee who performed R&D, thousand rubles (2020)	664,3	407,5
Number of advanced digital technologies developed (per 100,000 payroll employees) (2020)	0,9	7,3
Growth rate of the number of developed digital technologies (2016–2020), times	1,04	1,05
Investments in digital technologies per unit of financial result–income from the sale of goods, services, kopeks/rubles (2020)	0,86	1,48
Growth rate of investments in digital technologies (2016–2020), times	1,13	1,98
Share of investments in Russian (domestic) digital technologies in total investments in digital technologies (2020), %	54,8	31,5
Share of IT specialists in the total number of employees (2020), %	2,2	2,8
Number of advanced digital technologies used (per 100,000 employees) (2020)	249,1	896,6
Share of Russian advanced digital technologies in the total number of digital technologies used (2020), %	86,0	68,3

(continued)

Table 4 (continued)

Indicators	The value of the indicator in the holding "RZD"	Significance for the Russian economy
Growth rate of the volume of innovative goods, services produced and sold using digital technologies (2016–2020), times	1,24	1,30
Share of innovative goods, services produced and sold using digital technologies (2020), %	0,05	5,7

Compiled according to: [40, 46, 47]

[10, 11]; digital competition [14, 15]. At the same time, we are expanding the understanding of the "value" of a transport service that corresponds to the proper level of mobility. Early studies [48, 49] used just such an approach to the digital value of transport. In the analysis of the effectiveness of digital transport models, we based on research [50], which substantiated the dependence of the effectiveness of transport operators' operating models on costs reduced in the digitalization process. In understanding digital platform solutions, we have used the work on collaborative mobility modes [51], while significantly expanding their potential in relation to supply chain management platforms.

The most important debatable problem is the perception of the transport service digital value. Quite a lot of works are dedicated to the digital divide in the context of choice and access to a service, in particular, [52–54] and others. In our research, we specified the problem of the value of a transport service for a passenger and a cargo owner and developed platform solutions for open access to them.

5 Conclusion

Digitalization of Russian roads is the most important national project that creates new value propositions for the transportation of goods and passengers both within the country and in international transport corridors. It is the latter that forms the requirements for the quality of world-class transportation. To achieve this, the company has to change processes, technologies, management systems, operating models by activity and the entire business model, focusing it on value propositions for the customer.

Development of this research will continue in the direction of clarifying the values of the supply chain efficiency indices based on rail transportation (Logistic Performance Index), due to their digitalization, namely:

- Ease of arranging competitively priced shipments;
- Competence and quality of logistics services;
- Ability to track and trace consignments;
- Frequency with which shipments reach consignee within scheduled or expected time.

References

1. The World Bank. Logistics performance index: quality of trade and transport-related infrastructure. https://data.worldbank.org/indicator/LP.LPI.INFR.XQ?end=2018&start=2018&view=map. Accessed 21 Oct 2022
2. The World Bank. Logistics performance index (LPI). https://lpi.worldbank.org/. Accessed 21 Oct 2022
3. Connecting to Compete 2023 (2023) Trade Logistics in the Global Economy. The Logistics Performance Index and its Indicators. The International Bank for Reconstruction and Development/The World Bank. USA, Washington, DC., pp 90
4. Nalebuff BJ, Brandenburger AM (1996) Co-opetition. HarperCollinsBusiness, London
5. Nalebuff BJ, Brandenburger AM (1997) Co-opetition: competitive and cooperative business strategies for the digital economy. Strateg Leadersh 25(6):28–33
6. Armstrong JS, Clark T (1997) Review of Barry J Nalebuff and Adam M Brandenburger, Co-opetition 1. Revolutionary mindset that redefines competition and Co-operation. 2. The Game Theory Strategy that's Changing the Game of Business. https://repository.upenn.edu/marketing_papers/119. Accessed 01 June 2023
7. Cuofano G (2023) What is the value net model and Why it matters in business? FourWeekMBA. https://fourweekmba.com/value-net-model/. Accessed 22 May 2023
8. Daaboul J, Castagna P, Da Cunha C, Bernar A (2014) Value network modelling and simulation for strategic analysis: a discrete event simulation approach. Int J Prod Res 52(17):5002–5020
9. Woodruff RB (1997) Customer value: the next source for competitive advantage. J Acad Mark Sci 25(2):139–153
10. Allee V (2002) The future of knowledge, 1st edn. Routledge, London
11. Parolini C (1999) The value net: a tool for competitive strategy. Wiley, New York
12. Lanning MJ, Michaels EG (1988) A business is a value delivery system. McKinsey Staff Paper No 41. http://www.dpvgroup.com/wp-content/uploads/2009/11/1988-A-Business-is-a-VDS-McK-Staff-Ppr.pdf. Accessed 22 May 2023
13. Challenges for Competition Policy in a Digitalized Economy. European Parliament. Directorate-General for Internal Policies. Policy Department Economic and Scientific Policy A. (2015). https://www.europarl.europa.eu/RegData/etudes/STUD/2015/542235/IPOL_STU(2015)542235_EN.pdf. Accessed 06 June 2023
14. Parker G, Petropoulos G, Van Alstyne M (2020) Digital platforms and antitrust. Working Paper 06/2020, Bruegel
15. Parsheera S, Shah A, Bose A (2017) Competition issues in India's online economy. NIPFP Working Paper No. 194. https://doi.org/10.2139/ssrn.3045810
16. Baye MR, Prince J (2020) The economics of digital platforms: a guide for regulators. The Global Antitrust Institute Report on the Digital Economy, vol 34. https://doi.org/10.2139/ssrn.3733754
17. Zingales L, Lancieri FM (2019) Towards a democratic antitrust. Truth on the Market. https://truthonthemarket.com/2019/12/30/towards-a-democratic-antitrust/. Accessed 09 June 2023
18. Rusche C (2019) Data economy and antitrust regulation. Intereconomics 54(2):114–119. https://doi.org/10.1007/s10272-019-0804-5. (Springer, Heidelberg)

19. Shapiro C, Varian HR (2008) Information rules: a strategic guide to the network economy. In: Book in the academy of management review vol 30, no 2. Harvard Business School Press, Brighton. https://doi.org/10.2307/1183273

20. Bundeskartellamt (2017) Review of 2016 and future prospects. https://www.bundeskartellamt. de/SharedDocs/Publikation/EN/Pressemitteilungen/10_01_2017_Jahresr%C3%BCckblick_ EN.pdf?__blob=publicationFile&v=3. Accessed 13 June 2023

21. Jones CI, Christopher T (2020) Nonrivalry and the economics of data. Am Econ Rev 110(9):2819–2858. https://doi.org/10.1257/aer.20191330

22. Haftor DM, Costa Climent R, Lundström JE (2021) How machine learning activates data network effects in business models: theory advancement through an industrial case of promoting ecological sustainability. J Bus Res 131:196–205. https://doi.org/10.1016/j.jbusres. 2021.04.015

23. Carrière-Swallow Y, Haksar V (2021) Let's build a better data economy. The digital future. IMF. Finance and development. https://www.imf.org/en/Publications/fandd/issues/2021/03/how-to-build-a-better-data-economy-carricre. Accessed 09 June 2023

24. Haksar V, Carrière-Swallow Y, Giddings A, Islam E, Kao K, Kopp E, Quirós-Romero G (2021) Toward a global approach to data in the digital age. International Monetary Fund, pp 43. ISBN 9781513599427

25. Johnson MW, Christensen CM. Kagermann H (2008) Reinventing your business model. Harvard Business Review, Boston. https://hbr.org/2008/12/reinventing-your-business-model. Accessed 13 June 2023

26. Anthony SD (2012) Innovation is a discipline, Not a Cliché. Harvard Business Review, Brighton. https://hbr.org/2012/05/four-innovation-misconceptions. Accessed 15 Apr 2023

27. Furr N, Dyer J, Christensen CM (2014) The Innovator's method: bringing the lean start-up into your organization. Harvard Business Review Press, Boston, pp 288. ISBN 978-1625271464

28. Fera RA (2023) How Mondelez international innovates on the fly in 8 (Sort Of) easy steps. Fast Company. (02 July 2023). https://www.fastcompany.com/1682100/how-mondelez-intern ational-innovates-on-the-fly-in-8-sort-of-easy-steps. Accessed 15 Apr 2023

29. Transportation in Russia (2022) Statistical compendium. Rosstat. T. 65. M., pp 45–46

30. Container railway transportation in the Eurasian space in the first half of 2022. https://index1 520.com, date of the application: 01 Oct 2022

31. International Coordinating Council for Trans-Eurasian Transportation (CCTT). International Union of Railways (The UIC) https://icctt.com/mszhd, date of the application: 01 Oct 2022

32. Strategy for Digital Transformation of Russian Railways until 2025, approved. Board of Directors of Russian Railways 25 Oct 2019. https://ar2019.rzd.ru/pdf/ar/ru/performance-overview_ innovation-driven-development.pdf, date of the application: 01 Oct 2022

33. National Program and National Project "Digital Economy of the Russian Federation". https:// digital.gov.ru/ru/activity/directions/858, date of the application: 01 Oct 2022

34. Transport strategy of the Russian Federation until 2030 with a forecast for the period until 2035 (approved by the order of the Government of the Russian Federation dated November 27, 2021 No. 3363-p)

35. Departmental target program "Digital Platform of the Transport Complex of the Russian Federation", (approved by the Ministry of Transport of the Russian Federation on 28 Dec 2020)

36. Digital Transformation Program of the Ministry of Transport of the Russian Federation for 2021 and the planning period of 2022–2023, (approved by the Order of the Ministry of Transport of the Russian Federation dated 4 Feb 2021 № КБ-17-р)

37. Methodological recommendations for the digital transformation of public corporations and companies with state participation. https://digital.gov.ru/ru/documents/7342/, date of the application: 01 Oct 2022

38. The concept of the implementation of the integrated scientific and technical project "Digital Railway" 2017. M., 2017. https://www.zinref.ru/000_uchebniki/04600_raznie_2/173_cifrov aya_zheleznaya_doroga/000.htm?ysclid=livq8hl5c588917040, date of the application: 01 Oct 2022

39. Digital transformation: expectations and reality: reports for the XXIII Yasinskaya (April) International scientific conference on the problems of economic and social development, Moscow, 2022. (Abdrakhmanova GI, Vasilkovsky SA, Vishnevsky KO et al, and others; hands ed. count P. B. Rudnik; National Research University Higher School of Economics. M.: Publishing house of the Higher School of Economics (2022), pp 221)

40. Information on the development and (or) use of advanced production technologies: the results of statistical observation in f. No. 1 technology. https://www.consultant.ru/document/cons_doc_LAW_52009/ba78d400cb98182d76d5cc2e1f908b0124ebbb5d/, date of the application: 01 Oct 2022

41. Decree of the Government of the Russian Federation of 05/21/2022 N 931 (as amended on 03/25/2023) "On approval of the Rules for the exchange of electronic transportation documents and information contained in them between participants in information interaction, sending such documents and information to the state information system of electronic transportation documents, as well as the provision of other information related to the processing of such documents and information from the information system of electronic transportation documents to the state information system of electronic transportation documents at the request of the operator of the state information system of electronic transportation documents". https://consultant.ru, date of the application: 01 Oct 2022

42. Abdrakhmanova GI, Vasilkovsky SA, Vishnevsky KO, Gokhberg LM, et al (2022) Indicators of the digital economy. In: Statistical compendium. National research University "Higher School of Economics". M.: NRU HSE, 2023, pp 332

43. Methodology for calculating internal costs for the development of the digital economy: minutes of the meeting of the Government Commission for Digital Development dated September 27, 2019 No. 577 пр. https://rosstat.gov.ru date of the application: 01 Oct 2022

44. Presentation materials of the International Railway Salon space 1520 "PRO//Motion. Expo". https://railwayforum.ru, date of the application: 01 Feb 2023

45. Presentation materials, resolution of the XVII International Forum-Exhibition Transtec 2022. https://www.transtecforum.com, date of the application 01 Oct 2022

46. Information on the innovative activity of the organization: the results of statistical observation according to the form No. 4-innovation. https://rosstat.gov.ru, date of the application: 01 Oct 2022

47. Information on the use of information and communication technologies and the production of computer equipment, software and the provision of services in these areas: the results of statistical observation in the form No. 3-inform. https://rosstat.gov.ru, date of the application: 01 Oct 2022

48. Kaufmann V, Bergman MM, Joye D (2004) Motility: Mobility as capital. Int J Urban Reg Res 28(4):745–756. https://doi.org/10.1111/j.0309-1317.2004.00549.x

49. Kenyon S, Lyons G, Rafferty J (2002) Transport and social exclusion: Investigating the possibility of promoting inclusion through virtual mobility. J Transp Geogr 10(3):207–219. https://doi.org/10.1016/S0966-6923%2802%2900012-1

50. Davidsson P, Hajinasab B, Holmgren J, Jevinger Å, Persson JA (2016) The fourth wave of digitalization and public transport: opportunities and challenges. Sustainability 8(12):1248. https://doi.org/10.3390/su8121248

51. Boutueil V (2019) New mobility services. Chapter 3. In: Aguilera A, Boutueil V (eds) Urban mobility and the smartphone: transportation, travel behavior and public policy. Elsevier, Inc., Amsterdam, pp 39–78

52. Banister D (2019) Transport for all. Transp Rev 39(3):289–292. https://doi.org/10.1080/01441647.2019.158290

53. Hensher D, Mulley C, Ho C, Wong Y, Smith G, Nelson JD (2020) Understanding mobility as a service (MaaS): past, present and future, 1st edn. Elsevier Inc., Amsterdam

54. Lucas K (2019) A new evolution for transport-related social exclusion research? J Transp Geogr 81:102529. https://doi.org/10.1016/j.jtrangeo.2019.102529

Digital Transformation of the Corporate Accounting and Finance Process: Limitations and Risks for Russian Companies

N. V. Generalova, G. V. Soboleva, I. N. Guzov, S. A. Soboleva, and N. A. Polyakov

Abstract In this chapter, digitalization is considered from the perspective of the company in relation to the corporate business process "Accounting and Finance". The authors provide an overview of technologies preceding the digitalization of accounting and reporting, analyze the opportunities of digital technology implementation, and provide the digitalization experience of Russia's largest energy company, Rosseti. The chapter is based on published reviews and analytical reports prepared by authoritative audit companies, writing teams, and companies themselves. The authors identified the key blocks of digitalization constraints: the industry and scale of the company, the balance of benefits and costs, organizational structure readiness (for digitalization), and the lack of human resources; the following risks are identified: commercial risk, information security risk, responsibility segregation risk, legal and legislative risk, and risk of loss of process control. Based on the identified constraints and risks, a matrix of risks of the implementation of digitalization in the business processes of companies was made. This matrix is allow to build the architecture of digital transformation of the studied business process, which is determined by the internal constraints of a particular company.

Keywords Accounting in Russia · Big data · Blockchain · Digitalization in Russia · Digitalization of accounting · Digitalization of business · Digitalization

N. V. Generalova (✉) · G. V. Soboleva · I. N. Guzov (✉) · N. A. Polyakov
St. Petersburg State University, Universitetskaya nab. 7–9, 199034 St. Petersburg, Russian Federation
e-mail: n.v.generalova@spbu.ru

I. N. Guzov
e-mail: guzow@mail.ru

S. A. Soboleva
Technologies of Trust LLC, Grivtsova Lane, 4, 199000 St. Petersburg, Russian Federation

1 Introduction

In the last few years digitalization has become a widely discussed topic in the economic and social life of society. The debate is taking place at different levels: internationally, in the context of a single country, an industry, a company and individually. In Russia, at the state level, the Program of digital economy was adopted in 2017, designed for 2017–2030 [1]. An analysis of the indicators of digitalization of business in Russia reveals a certain lag in this sphere from the leading countries. For example, specialists of the Higher School of Economics (HSE) developed the index of digitalization of business which measures the speed of adaptation of companies to digital transformation and describes the use of digital technology. For Russia it is 28 points, which puts Russia next to such countries as Bulgaria, Hungary and Romania [2]. At the same time, the leaders in the level of digital technology in the business sector have a significantly higher value of this index: Finland (50 points), Belgium (47), Denmark (46), the Republic of Korea (45). At the same time, Russian business has a leading position in some areas of digitalization. For example, in the application of technology in the banking sector Russia is among the top five leading countries in Europe in terms of digital banking development [3]. McKinsey's study showed that the leading banks in Russia perform 1.5–2 times more transactions than the largest European banks when they provide mobile applications to their customers. Depending on the service method, 58% of customers use remote banking (of these, 15% use only online banking, 10% use only mobile banking, and 32% use both platforms) [4].

According to research by Rosstat and HSE, Russian companies have already widely adopted basic and relatively simple digital technologies (83% in 2017 and 86.6% in 2019 of Russian organizations already use broadband Internet, 63% in 2017 and 68,7% in 2019 have adopted electronic data exchange technologies); a smaller proportion of companies have carried out deep automatization and restructured their business processes for advanced digital technologies (cloud services–23% in 2017 vs. 28.1% in 2019, ERP systems–12.2% in 2017 vs. 14.8% in 2019, RFID technologies–5% in 2017 vs. 6.7% in 2019) [5, 6]. At the same time, experts highlight that the digitalization of Russian manufacturing enterprises remains at a low level compared to foreign competitors. The use of such technologies as computer engineering, virtual modeling, additive technologies, mechatronics and robotics has not yet become widespread in Russian business [2].

From the perspective of the company, digital change is usually based on the BEOM (Business Entity Ontological Model), which allows for systematic structuring and describing activities by tasks, organizational structures, territories and objects, organizing and translating its experience accumulated in specific situations throughout the life cycle. In the context of digitalization, the enterprise is "decomposed" into business processes: operational (technological) and corporate, for each of which digital technologies and the effects of their use are prescribed.

Most researchers highlight the benefits that will emerge from the implementation of digital technologies: transparency, accessibility and relevance of financial information to external users, the obviousness of the way of collecting, managing and analyzing financial data within the company, optimization of decision-making, the quality of financial reporting, the usefulness of accounting information and the effectiveness of strategic decision-making [7–9].

The digital transformation of the economy affects all areas of life, including the economy and business. From a business perspective, both operational (technological) and corporate processes are undergoing changes due to digitalization or digitization. The former–operational (technological) processes–are caused mainly by the industry and the specifics of the company's activity. The latter–corporate processes–are relatively common for companies and include investment activities, risk management, personnel management, legal support, logistics, finance and accounting etc. Technological changes of corporate process "Accounting and Finance" should be considered in the context of transformation of business model of the company as a whole, taking into account external and internal risks and limitations.

2   Research Methodology

In this study it was not intended to introduce new terms and concepts to those already used in the professional literature, practice, thematic reports and reviews, although widely discussed. Among them are digitalization, digital transformation, digital economy, digitalization of business, digital environment, digital ecosystem, etc. We share the position of experts that there are "shades" of terminology and terms should be applied in the right context to convey the correct meaning in order to avoid semantic inconsistencies and the search for harmonized definitions of key concepts of digitalization is not over [2, 10].

With regard to the digitalization of the corporate Accounting and Finance process, let us draw attention to the distinction between the concepts of "automation" and "digitalization" in accounting. Both of these processes are intended to replace manual labor. The fundamental difference between automation is that automation is a substitute for the way information is entered and processed, without significant opportunities for its further use and exchange. Digitalization, based on the information entered in electronic form, is aimed at its further transfer into the general information system of the company for its further sharing, generating on its basis new data and decisions, including those made by this digital system itself, providing various participants in the business processes of the company and external users of information with not only information, but also broad opportunities to use this information. Thus, accounting automation is the predecessor of digitalization of accounting and reporting; and now automation is one of the approaches to the technical organization of accounting and reporting, in other words–digitalization uses automation as one of the tools, but not technologies [11].

This study is based on the study of available reviews and analytical reports performed by major international audit companies, authoritative author teams and companies themselves, as well as verification of the data obtained with the help of

Russian companies' reports on digitalization. The purpose of this study is to identify the role of digitalization of the corporate Accounting and Finance process in
the digitalization of the company as a whole, as well as to identify the limitations
and risks encountered in the implementation of digital technologies. In this regard,
it is important to analyze the applied digitalization technologies in accounting and
reporting in order to identify the factors that influence their variability, including
the correlation of the set of technologies applied in the "Accounting and Finance"
process with the overall set of digital technologies as applied to the company as a
whole.

The authors formulated two key hypotheses.

Hypothesis 1. The transition to digital technologies in the corporate process
"Accounting and Finance" is significantly influenced by both internal and external
factors, but the set of digital technologies is much narrower than the digitalization
of the business as a whole.

Hypothesis 2. The existing risks and limitations of digital technologies cannot
allow uniform approaches to the digitalization of the business process "Accounting
and Finance" for all business processes participants. The architecture of the digital
transformation of the studied business process will be determined by the internal
constraints of a particular company.

3 Technologies of Digitalization of the Corporate Process "Accounting and Finance"

Digitalization of accounting and financial reporting in companies is not the first step
in improving the technologies used in this area. For the most part, the transition
to machine processing of information arrays has already taken place. This section
provides an overview of the technologies that are already in use and are widely
used not only by large Russian companies, but also by medium and small ones
(pre-digitalization technologies), as well as technological solutions in the area of
accounting that have emerged in the conditions of digital transformation (digital
technologies in accounting and reporting).

3.1 Pre-digital Accounting Technologies

The use of electronic tables (Microsoft Excel, Access, etc.) for accounting and
reporting purposes made it possible to identify and systematize the available information according to the required criteria, perform its criterion selection, form separate
tables, derive results, combine information and bring it together in a single reporting
document. At the same time, this method of automating reporting has certain risks:
missing or making mistakes; loss of data when several users work; difficulties in data

audit (entails time and labor costs), etc. Nevertheless, Excel tables are still widely used as a reporting transformation tool. For example, Russian companies use this "traditional" tool for the releasing of consolidated financial statements under International Financial Reporting Standards (IFRS), making adjustments to the uploaded accounting data under Russian Accounting Standards (RAS) in order to bring them in line with IFRS [12].

Application of specialized software products ("1C: Enterprise", "BEST", "Parus", etc.) allows to carry out key functional set of (financial) accounting of the company (input of primary documents, registration of facts of economic life of the organization, logging of business transactions), on the basis of this information to form accounting, tax and management reports. Programs need constant updates of entered data, which entails significant time costs. Despite the fact that specialized products can be adapted to the specifics of any company, independent transformation of processes in programs for unqualified staff is significantly limited. A competitive disadvantage of these software products in front of their successors will be the non-provision of forecasting risks and probable outcomes of implementation of various strategic alternatives of investment decisions, lack of tools for the formation of a variety of reports for corporate governance purposes.

Integrated ERP-systems of enterprise management (Enterprise Resource Planning) provide the possibility of combining several subject areas of automation. Financial accounting is integrated with subsystems of operational and production accounting, production planning, budgeting, etc. The main advantage of ERP-systems is the ability to solve several problems simultaneously: planning and accounting of production and logistics operations, rationing production costs, cost formation and assessment of the effectiveness of the company. ERP-systems provide management with operational information on various aspects of the company's activities, promote transparency of the process of planning and control, allow to model alternative scenarios of the company's development. Among the disadvantages of ERP-systems we note the high cost of licenses, the duration of implementation, the necessity for qualified and well-paid IT professionals who will be able to configure and administer the system. According to a study conducted in 2018 among companies based in Germany by PWC experts, the most common ERP software products are the following four: SAP (61%), PeopleSoft (24%), Oracle (5%), Microsoft Navision/AX (5%); the rest occupy insignificant shares: JD Edwards (2%), Sage (2%), and others (1%) [13]. In Russia, a domestic product is widespread–1C: ERP Enterprise Management, which has a similar functionality, but is distinguished by a significantly lower cost and adaptability to the Russian reporting forms.

Electronic systems–digital accounting. Modern accounting systems use electronic systems more and more: electronic document management system, ECM-system (Enterprise Content Management), the system of electronic delivery of reports, etc. Electronic digital signature is an indispensable attribute for the implementation of electronic document flow.

3.2   Digital Technologies Used in Accounting and Reporting

There is no consensus on the exact set of digitalization technologies, as their classification varies, and as more and more new technological solutions emerge, leading to additions to the list of digitalization tools. For example, the 2019 Digitalization Study in Russia highlighted eight of the most popular digital technologies for implementation, indicating the percentage of use by respondent companies: 1. big data analysis (Big Data) and predictive analytics–68%,. chatbots–51%, 3. robot process automation (RPA) office processes–50%, 4. Optical recognition (OCR)- 36%, 5. artificial intelligence (AI)–28%, 6. internet of things (IoT)–24%, 7. virtual reality (VR/AR)–21%, 8. blockchain–19% [14]. This toolkit is the most widely used, according to a survey of more than 100 Russian companies from various industries, including the financial, telecommunications, metallurgy, oil and gas, transport and retail sectors. We emphasize that this list of technologies reflects the digitalization of all business processes in the company, not only the digitalization of the accounting structure. Moreover, according to our estimates, these percentages of application of digitalization tools reflect the results of technology use by major Russian companies rather than represent the realities of medium and small businesses. That is so especially because "application" includes not only the implementation of technologies, but the testing and pilot project phase.

The set of digital technologies used in the accounting business process will be different from the set used for all company processes, technological and corporate. Analyzing the few studies on the applied technologies in accounting and reporting, it can be stated that the set of digital solutions for accounting is significantly narrower. Here are some examples. A large-scale study on digitalization in accounting and reporting conducted among more than 100 German companies in 2017 presents a set of 11 digital solutions with the percentage of use among respondent companies, and these "newest digital capabilities, are designed to automate routine accounting processes as much as possible." 1. uniformity of systems–64%, 2. management of data quality–56%, 3. paperless accounting, 9. interfaces to (external) systems–18%, 4. integrated consolidation system–44%, 5. creation of transparency–33%, 6. process automation–25%, 7. big data analyses–23%, 8. real-time reporting–19%, 10. tools for visualization–16%, 11. cloud computing–7% [15]. The scale of application is significant, although also characteristic of large companies, including such as Anonymous, Siemens AG, BMW AG. Robots, chatbots and the Internet of Things are not reflected in this list of accounting and reporting solutions. A profile article by Croatian colleagues published in 2019 identified four other key accounting technology solutions "with potential and feasibility for adoption." 1. artificial intelligence, 2. blockchain, 3. continuous accounting, and 4. big data [16]. The above lists of applied technologies and technological solutions largely reflect the retro view of five years ago, and obviously it has undergone changes in the rapid development and spread of technology, taking into account the impact of the pandemic caused by Covid-19.

A different approach to the development of digitalization of accounting can be demonstrated by taking the hierarchical structure of the accounting process (accounting process pyramid) and designating digital technologies for each element: 1. primary documents describing facts of economic life–mirrors/digital twins, 2. accounting registers–robotization, 3. financial statements–blockchain [11, 17].

XBRL reporting and reporting taxonomy. Digitalization technologies have not only affected the accounting process, they are applied at the stage of financial reporting indicators formation. XBRL (eXtensible Business Reporting Language) technology, which originated in the United States, is a tool of communication and exchange of business information between systems. XBRL initiative was supported in many countries by stock exchanges and securities market regulators and bank regulators. Communications are based on metadata sets (data about data) that contain description of both individual reporting indicators and relationships between them and other semantic elements of taxonomies. XBRL taxonomies contain definitions and characteristics of individual reporting elements as well as characteristics of relationships between these elements, which allow to implement a range of operations with reporting data, provide an automatic link to regulatory requirements and various reporting standards. Based on the taxonomy, organizations create a package of reporting data (figures, text and graphic elements linked to the taxonomy). In Russia, the introduction of XBRL reporting began in 2015 with the creation of a Russian XBRL jurisdiction made by Central Bank of Russia (CBR). According to the CBR's requirements, starting in 2018 this reporting format became mandatory for insurers, professional securities market participants, joint-stock investment funds, management companies of investment funds, mutual funds and non-governmental pension funds. In 2021, brokers, credit rating agencies and specialized depositories will switch to the XBRL format, and after 2021–credit history bureaus [18]. The advantages of XBRL reporting from the perspective of investors and consumers are the simplicity and convenience of analytical tools based on portal solutions, as well as machine-readable information disclosure understandable to international investors. The financial market, supervisory authorities and federal executive bodies, in addition to investors, will benefit from such a technical solution for the presentation of reports.

Technologies of corporate reporting presentation. In the context of accounting procedures for reporting, it should be noted that one of the key trends in the recent history of public reporting of companies is its separation into an independent significant accounting procedure [19]. Moreover, the demand for financial statements of public companies has increased significantly due to the rapid increase of users who have gained access to investing in the securities of companies by opening brokerage and investment accounts. The idea of creating an interactive multidimensional model of financial statements is to proactively find a solution that will allow each user to have a significant flexibility in providing data: for example, depending on the target, the user will be able to group and match income and expenses, getting the required option for calculating the financial result, include or not include in the balance sheet assets that do not belong to the organization but are under management (e.g., leased assets), etc. Interactive multidimensional model of corporate reporting, which includes financial and non-financial indicators, will promptly detail, aggregate and restructure the initial form of providing reporting data according to the user request. The current trend in the development of interactive reporting is its transformation into a special database, or rather a special class of data marts. Users who systematically study the reporting prefer to convert it into a database, and use it to obtain reports of different forms at their discretion. The creation of and subsequent work with the database allows to simplify the procedure of storing, processing and subsequent presentation of information in the most differentiated ways.

Summing up the review of technologies and technological solutions in accounting and reporting, it can be stated that the replacement of manual labor in certain areas (automation) is being replaced by a new model of digitalization of accounting processes, which has fundamentally different capabilities and meets the current challenges of modern society. Networked data processing and multimedia accounting performed by robots and artificial intelligence, and as a result, visualized reporting capable of adapting to the request of a particular user–this is if not the present, then foreseeable future.

3.3 Cases of Digitalization of Russian Companies: ROSSETI and Others

In the context of the issues under consideration, the experience of Russian electric grid company Rosseti PJSC, which in 2018 published its Digital Transformation 2030 Concept, which formulated "the main directions of technological and organizational changes in the company's work to find new mechanisms, methods, algorithms of corporate and technological management of the company's processes and their subsequent transformation to improve the efficiency and quality of services provided and their accessibility" [20]. In the analyzed Concept it is noted that the anticipated digital changes are based on an ontological model of activity, taking into account the requirements of the network-centric approach, which reflects current trends in the implementation of digital transformation in the electric power industry. The developers of the Concept of ROSSETI operate with Industry 4.0 technologies, the list of which and the anticipated effect of their implementation are presented in Table 1.

A comparison of the technologies planned to be implemented by ROSSETI with the set of "new technologies" presented by experts according to a survey of 40 leading global energy companies, shows that the Russian electric power leader has included all technologies used by industry companies [21].

Digital management of the company is designed to unite a set of corporate information systems and, as a result, to ensure transparent, reliable, automated data exchange, eliminating duplication of information and manual data entry. A prerequisite for ensuring the reliability of operational management information is the use of a single data model of technological and business processes that unambiguously defines the essence of these processes and the relationship between them for all Rosseti subsidiaries and dependent companies, which is important in the context of generating consolidated reporting. In terms of the digitalization of accounting and

Table 1 Rosseti's digital technologies and the expected effect of their application

№	Technology	Possible impact and effect of implementation on company processes
1	Business Ontology	Gradual digitalization (optimization) of the company's core business processes. Reduction of the cost of all business processes of the company
2	Digital Shadows	As part of the development of online and offline decision support systems, creation of mathematical models of networks, objects, processes, etc. Reducing operating costs and developing new business for the company
3	Industrial Internet of Things (IoT)	Significant CAPEX and OPEX reductions for collecting data from remote objects and devices on the network, including a qualitative increase in the volume of this data. Reduced operating costs and development of new business for the company
4	Big Data	Significant increase in the transparency of operations, qualitative saturation of online and offline decision support systems with data. Optimized decision-making on the operational and prospective environment. Additional effects due to overall processing of technological and corporate data
5	Machine learning	Automated processing of data sets within online and offline decision support systems tasks with appropriate mathematical algorithms. Optimized decision-making for operational and prospective activities
6	Blockchain	Elimination of intermediaries in the kWh sales chain to the final consumer, transition to automated Smart contracts, service development for active consumers and distributed energy. Development of new types of services (business) of grid companies for market participants
7	Visual perception and decision-making	Increased reliability. Ability to predictively of the appearance of a threat to of disconnection of power grid equipment
8	Artificial intelligence	Increased reliability. Predictive reporting of the threat of power grind outage
9	Remote scanning for creating 3D models of network elements	Improved capital and operating cost efficiency (OPEX/CAPEX). Increased adaptability
10	Virtual reality (simulation of 3D images or full environment), Augmented reality	CAPEX reduction. Increase of adaptability. Creation a system for control over the implementation of investment programs in an automated mode

Source Compiled by the authors based on ROSSETI's Digital Transformation 2030 Concept, https://www.rosseti.ru/investment/Kontseptsiya_Tsifrovaya_transformatsiya_2030.pdf [20]

reporting, Concept 2030 stipulates the use of only one technology, Big Data, for the Finance, Economics, and Accounting process as one of the company's corporate processes and is designed to produce three key effects: automated reporting, a system for monitoring the implementation of business plans and increased adaptability. Thus, despite the large choice of digital technologies presented in Table 1, one of the largest Russian companies does not plan to widely use all the possibilities for accounting processes, but focuses on those that, in its opinion, will allow to achieve the set objectives. At the same time, for the digitalization of "risk management" business process, it plans to additionally use artificial intelligence technology, which is designed to facilitate the development of recommendations on risk leveling.

The analysis of digitalization of Russian companies has certain difficulties due to the fact that the availability of information on public resources is very limited. In this regard, of particular interest is the analytical report, carried out by domestic experts published in 2021 [22]. The leading companies in digital transformation, whose experience was studied according to the annual reports, are the following: Russian Railways, Gazprom Neft, Rosseti, and Lukoil. These companies have detailed digital development strategies for the long term up to 2024–2030. Comparing the strategic documents of these four companies, we note some common features inherent in the digital strategies of corporations: (1) the main objective of the strategy is to improve company efficiency and the quality of services provided to customers, (2) digital transformation is accompanied by process reorganization and optimization, (3) a necessary element of digitalization is the creation of competence centers and various training and experimental structures ("factories"), (4) the most popular trends are the creation of digital twins, the use of artificial intelligence and work with big data, (5) strategies describe mass participation in production chains of "digital worker", equipped with augmented reality (AR) gadgets and computer software for planning and control of operations. Information on the applicable digital technologies by the analyzed Russian companies is presented in Table 2.

4   Limitations and Risks of Digitalization of the Business Process "Accounting and Finance"

Digital technology not only creates benefits, but also generates risks and has limitations. This study focuses on digitalization of accounting and reporting as an infrastructure component of corporate governance. The limitations can be broken down into several large blocks that reflect the most important challenges in implementing digitalization. The risks are also grouped into areas.

Table 2 Comparison of digital strategies of Russian companies RZD (Russian Railways), Gazprom Neft, Lukoil and Rosseti

Companies	Facts about strategy	Trends and/or technologies	Tools and approaches
RZD (Russian Railways)	Adopted 29.10.2019, until 2025	Internet of Things, big data, distributed registries, machine learning, VR and AR, quantum communications	Digital platforms, Reorganization of business processes, introduction of cross-functional interaction mechanisms, creation of an institute of "change agents"
Gazprom Neft	Adopted 16.09.2019, set to 2030	AI, VR and AR, unmanned aircraft systems, video analytics, tech vision, smart devices, blockchain, robotic processes	Competence centers and IT clusters, technology centers, digital technology vision, corporate cloud approvals, regulations and guidelines, corporate training system
Lukoil	Adopted in 2018, set to 2027	Digital twins, machine learning, digital platforms, AI, robotics, digital predictive analytics	Integrated modeling, neural networks
Rosseti	Adopted 21.12.2018, set to 2030	Data Factory, digital network, analytics competence center, cybersecurity center	AI research center, internet of things, big data, digital twins, 3D models, VR and AR, distributed registry

Source Compiled by the authors on Bertyakov et al. [22], p. 184

4.1 Limitations

Industry, product, scale. The corporate "accounting and reporting" process is influenced by technological processes, which in turn are conditioned by industry affiliation, the specificity of the company's activity, scale and other factors. According to the research, the level of digitalization in the company depends on its size (the larger the company, the higher the degree of digitalization), and on the industry (the more technological the company's product, the higher the degree of digitalization) [14, 23]. The set of digital technologies used in accounting processes is significantly lower than for all processes [24]. This can be clearly seen in the above example of Rosseti PJSC: for the business process "Finance, Economics and Accounting," only one technology–Big Data–is provided. The prospects of using distributed registers technology (blockchain) in accounting and auditing are actively discussed [11, 17, 25].

Balance of benefits and costs. This is a classic limitation that is embedded in many technical standards. For example, in IFRS it is allowed not to apply accounting techniques if the benefits from their use are lower than the costs of their implementation. In the digital context, this restriction has two aspects. On the one hand, this is the company's point of view: its overhead costs for the implementation of a particular digital technology must be covered by its own revenues. In this case, an additional difficulty is that the implemented technology will not always be scalable and "boxed" solutions will not always be suitable. The second aspect is the cost-effectiveness in the context of the concept of sustainability and ESG. For example, blockchain technology is associated with high energy consumption, which at least raises the question of its energy efficiency in the context of environmental protection.

Organizational readiness. The need for structural preparation of the company for the application of digital technology is an important element in the success of the implementation of new technologies. For example, the robotization of company accounting processes is only possible if they are structured, repeatable, based on uniform rules and controlled by digital data input [26]. As another example, the security benefits of blockchain, which make it supposedly immutable, in an accounting environment due to the high number of business transactions on the external loops of the system, make data not fully accessible or reliable [27].

Human Resources. There is a shortage of qualified personnel. There will be a need for 'techies' and 'accounting techies' who can work with digital data and digital software. This entails a restructuring of training and professional development. And despite the fact that universities are actively introducing new courses on digitalization, the labor market is clearly not yet ready to meet the demand for an "accountant in the digital economy [28, 29].

4.2 Risks

Commercial Risk. The use of digital technology in accounting can lead to information containing trade secrets being lost or, because of openness, being used to the detriment of the company by competitors or regulators.

Information security risk. This negative side of digital adoption is mentioned in almost all studies. For example, in a 2019 KPMG study, more than half of Russian respondents noted this "threat of digitalization" [14]. The budgets of companies to combat cyber-fraud are growing year after year. The importance of this group of risks is confirmed by the increasing cybercrime: hacking and attacks on mobile devices and financial mobile applications as part of the infrastructure of remote banking and payment systems, attacks on smart contracts, etc. It is important that this group of risks is not only of external, but also of internal nature.

Risk of responsibility delimitation. Introduction of digital accounting and financial procedures inevitably entails automation of decision-making. In the case of violations, errors occurring in digital processes, the personalization of responsibility may be lost or simply not defined from the outset. The example of liability in accidents involving unmanned (highly automated) vehicles is relevant. With regard to accounting, who will be responsible if a robot makes a decision based on a preset algorithm, which in the rapidly changing realities will result in losses for the company?

Legal and legislative risk. Currently, there is a significant legal vacuum in the use of digital technology, because current Russian legislation does not contain comprehensive rules, and the bylaws do not contain clear and understandable mechanisms for interaction. In particular, because of this, a company may be denied the recognition of expenses for tax purposes, because the "documentary" evidence will be outside of the current legal framework. This also includes the risk of loss of property, risk of copyright infringement and so on.

The risk of loss of control over process management. If a company uses "boxed" products and cloud technologies, it runs the risk of losing control, because process algorithms are set up outside the company. These risks are the higher, the smaller scale of business is and the less opportunity there is to influence the vendor of the product. The result may be the deterioration of process controllability and efficiency in the short term, and in critical cases catastrophic losses may occur.

4.3 Risk Assessment Matrix for the Implementation of Digitalization in the Business Processes of Companies

Based on the identified constraints, it is proposed to assess the risks of implementing digitalization in the business processes of companies through a digital risk assessment matrix. This matrix should be used in relation to the profile of a particular organization when implementing the processes of digitalization of business for the selection of digital technologies. Table 3 shows the digitalization risk assessment matrix for the profile of "large high-tech company".

Table 3 Risk assessment matrix of digitalization processes in the example of a large high-tech company

Limitations	Risks				
	Commercial risk	Information security	Delimitation of liability	Legislative risk	Risk of loss of process control
Industry, product, scale	M	M	H	H	M
Balance of benefits and costs	L	L	M	M	H
Readiness of organizational structure	L	L	L	M	M
Human resources	M	M	L	H	H

L–low risk, M–medium risk, H–high risk
Source Compiled by the authors

5   Conclusion

This study was aimed at identifying the relationship between the digitalization of business in general and the digitalization of the corporate business process "Accounting and Finance" as applied to Russian companies. The scale and pace of digitalization of the accounting component of business, in our opinion, reflects the realities, and sometimes only the desired picture of survey respondents - representatives of large and major Russian and international companies. Medium and small businesses will "digitize" much later and in a more truncated palette of technological solutions. The proposed digitalization technologies must be tested, including their environmental impact, due to the energy they consume. Otherwise, the intended effect of digitalization will be offset by the damage that will be done to the human environment. In particular, the energy efficiency of blockchain is controversial. However, the undeniable benefits of digitalization technologies are significant. In accounting and reporting, digitalization will lead to the replacement of colossal manual labor, with intelligent, standardized "labor" of smart by systems that can record and process large amounts of data. Company performance indicators (both financial and non-financial) will be generated through digital technologies, mainly by the XBRL taxonomy, and there will be data from one synchronized integrated system, rather than "unknowable" amounts from different departments. As a result of such changes, information from financial accounting and financial statements will be increasingly possible to process and broadcast in other resources (e.g., brokerage accounts, statistical databases, analytical reviews, etc.). This will increase the demand for accounting data. Digitalization will allow to implement the idea when public reporting will be created by users according to their individual needs on the basis of a single array of data certified by the auditor.

The hypotheses proposed by the authors have generally been confirmed. Regarding Hypothesis 1, the analysis of studies and reports on the progress of digitalization, as well as the case of Rossetti PJSC, confirmed that the list of digital technologies applied in the corporate process "Accounting and Finance" is much narrower than the set of digital technology solutions applied in other areas of business digitalization. Big Data technology is the most common digital technology in accounting and reporting. To this day, ERP systems are widespread in Russian companies. Such technologies as blockchain, barcoding, machine vision, augmented reality, and robotization are used to a lesser extent. The author's hypothesis 2, that the existing risks and limitations of digital technology application cannot allow for uniform approaches to the digitalization of the business process "Accounting and Finance" for all participants of business processes, according to the analysis of reports and reviews, was generally confirmed. The four main blocks of constraints identified by the authors: industry affiliation and scale of the company, balance of benefits and costs, readiness of the organizational structure, lack of human resources, cover the defining indicators when choosing digital solutions for the business process "Accounting and Finance". The following risks have been identified: commercial risk, information security risk, risk of delimitation of responsibility, legal and legislative risk, and risk of loss of process control. As a follow-up to Hypothesis 2, a matrix of digitalization constraints and risks was proposed, which can be used to assess the risks of implementing digital processes at the company level. At the same time, the application of this matrix requires additional research and analysis of the progress of digitalization in companies. A further direction of research is the development of a matrix to assess the risks of digitalization, based on the identified constraints and risks for the profiles of companies with a set of recommendations for the application of digital technologies.

Currently there are qualitative changes in the application of technology, when the previously existing processes of data transfer and processing are not accelerated, but fundamentally new technologies are emerging. These technologies will lead to a change in the role and importance of the scientific component in the system of risk-oriented management of the company. And thanks to technologies such as big data, blockchain, artificial intelligence and others, the financial accounting and reporting service will be able to establish itself as an equal component of companies' business processes.

The authors are by no means opposed to digitalization in general and digitalization in accounting in particular. On the contrary, by discussing the "thin spots", optimal approaches to a digital strategy can be developed. This paper identifies the limitations and risks that are currently being highlighted by both researchers and companies.

References

1. Program Digital Economy of the Russian Federation. Decree of the Government of the Russian Federation of 28.07.2017 N1632-r. http://government.ru/docs/28653/. Accessed 24 July 2023
2. What is the digital economy? Trends, Competencies, Dimension XX–Moscow. https://publications.hse.ru/chapters/290233040. Accessed 24 July 2023
3. Petrova LA, Kuznetsova TE (2020) Digitalization in the banking industry: digital transformation of environment and business processes. Financ J 13(3):91–101. https://doi.org/10.31107/2075-1990-2020-3-91-101
4. Shaikh AA, Karjaluoto H (2019) Marketing and mobile financial services: a global perspective on digital banking consumer behavior, 1st edn. Routledge
5. Information Society: the main characteristics of the subjects of the Russian Federation, https://rosstat.gov.ru/storage/mediabank/info-ob_reg2018.pdf. Accessed 24 July 2023
6. Information Society in the Russian Federation 2020. https://rosstat.gov.ru/storage/mediabank/lqv3T0Rk/info-ob2020.pdf. Accessed 24 July 2023
7. Moll J, Yigitbasioglu O (2019) The role of internet-related technologies in shaping the work of accountants: new directions for accounting research. Br Account Rev 51(6). https://doi.org/10.1016/j.bar.2019.04.002
8. Phornlaphatrachakorn K, Kalasindhu KN (2021) Digital accounting, financial reporting quality and digital transformation: evidence from Thai listed firms. J Asian Financ Econ Bus 8(8):409–419. https://doi.org/10.13106/jafeb.2021.vol8.no8.0409
9. Schiavi GS, Momo F da S, Behr, A, Maçada ACG (2020) On the path to innovation: analysis of accounting companies' innovation capabilities in digital technologies. Rev Bus Manag 22(2):381–405. https://doi.org/10.7819/rbgn.v22i2.4051

10. Tsenzharik MK, Krylova YV, Steshenko VI (2020) Digital transformation in companies: strategic analysis, drivers and models. St Petersburg Univ J Econ Stud 36(3):390–420. https://doi.org/10.21638/spbu05.2020.303
11. Generalova NV, Guzov IN, Soboleva GV (2021) Digitalization of accounting and auditing: evolution of technology, Russian experience and development prospects. Financ Bus 17(4):63–80. https://doi.org/10.31085/1814-4802-2021-17-4-112-63-80
12. Sokolov, V. I., Generalova, N. V., Guzov, I. N., & Karelskaia, S. N. Applying IFRS in Russia. In P. Weetman & I. Tsalavoutas (Eds.), The Routledge companion to accounting in emerging economies, pp. 42–55. Routledge. (2019). ISBN 9781351128506
13. Digitalisation in finance and accounting and what it means for financial statement audits. PWC (2018)
14. Digital Technologies in Russian Companies. https://roscongress.org/en/materials/tsifrovye-tekhnologii-v-rossiyskikh-kompaniyakh/. Accessed 24 July 2023
15. Digitalization in Accounting Study 2021. https://kpmg.com/de/de/home/themen/2021/09/digitalisierung-im-rechnungswesen-2021.html. Accessed 24 July 2023
16. Gulin D, Hladika M, Valenta I (2019) Digitalization and the challenges for the accounting profession. In: Proceedings of the ENTRENOVA-ENTerprise REsearch InNOVAtion Conference, Rovinj, Croatia, 12–14 September 2019, IRENET-Society for advancing innovation and research in economy, Zagreb, vol 5, pp 502–511
17. Titov V, Shust P, Dostov V, Leonova A, Krivoruchko S, Lvova N, Guzov I, Vashchuk A, Pokrovskaia N, Braginets A et al (2022) Digital transformation of signatures: suggesting functional symmetry approach for loan agreements. Computation 10(7):106. https://doi.org/10.3390/computation10070106
18. XBRL open reporting standard. https://cbr.ru/projects_xbrl/. Accessed 24 July 2023
19. Razletovskaia V, Stepnov I, Guzov I, Svetlichnyy S (2023) Development of modern financial technologies at the national and international levels. In: Ilin I, Petrova MM, Kudryavtseva T (eds) Digital transformation on manufacturing, infrastructure & service. DTMIS 2022. Lecture notes in networks and systems, vol 684. Springer, Cham. https://doi.org/10.1007/978-3-031-32719-3_23
20. Rosseti Concept Digital Transformation 2030. https://www.rosseti.ru/upload/iblock/582/rajp59pvuvjsx5ztr38jjz2q98o8rkbd/Kontseptsiya_Tsifrovaya_transformatsiya_2030.pdf. Accessed 24 July 2023
21. 2019 Digital operations study for energy. https://www.strategyand.pwc.com/gx/en/insights/2019/2019-digital-operations-study-for-energy/2019-digital-operations-energy-insights.pdf. Accessed 24 July 2023
22. Bertyakov AV, Vinichenko OA, Vitushkin VA et al (2021) Digital transformation strategy: write to execute. RANEPA, Moscow
23. Digital Transformation in Russia 2020 KMDA. https://komanda-a.pro/projects/dtr_2020. Accessed 24 July 2023
24. Kreher M, Sellhor T, Hess T (2017) Digitalisation in accounting-study of the status Quo in German Companies, KPMG
25. Prusova VI, Kolokolova EY, Zhidkova MA, Kargina AV (2021) Diffusion of digitalization. In: The customs sphere through electronic document management. intelligent technologies and electronic devices in vehicle and road transport complex TIRVED 2021. Moscow, Russian Federation, pp 1–4. https://doi.org/10.1109/TIRVED53476.2021.9639101
26. Kokina J, Blanchette S (2019) Early evidence of digital labor in accounting: innovation with robotic process automation. Int J Account Inf Syst 35, Article 100431. https://doi.org/10.1016/j.accinf.2019.100431
27. Coyne JG, McMickle PL (2017) Can blockchains serve an accounting purpose? J Emerg Technol Account 14(2):101–111. https://doi.org/10.2308/jeta-51910
28. Soboleva GV, Zuga EI (2022) The participation of Russian companies in the implementation of the ESG agenda: Social and corporate aspects in the context of non-financial reporting. St Petersburg Univ J Econ Stud 38(3):365–384. https://doi.org/10.21638/spbu05.2022.302

29. Guzov IN, Polyakov NA, Guzov YI (2022) Auditors in Russia: entry into the profession. In book: Challenges and solutions in the digital economy and finance. In: Proceedings of the 5th international scientific conference on digital economy and finances DEFIN 2022. Springer Proceedings in Business and Economics, St. Petersburg, pp. 517–532. https://doi.org/10.1007/978-3-031-14410-3_54

New Opportunities for Predicting Heat Supply to Consumers in the Context of Digitalization of Operation Processes

Natalia G. Verstina⬤, Olga F. Tsuverkalova⬤, and Nikolay A. Verstin⬤

Abstract The article presents the results of a study of the activities of heat supply organizations that get the opportunity to improve business efficiency through the introduction of digitalization technologies in the context of sustainable development goals (SDGs). Based on a preliminary analysis of the activities of heat supply organizations for a reference group of countries using district heating supply to consumers, a problem area that needs digitalization has been identified - the operation of heat networks from the source to consumers. As research materials, data on the heat supply of consumers were used, which showed new possibilities provided by digitalization for predicting the processes of operation of heat networks based on the values of the operating parameters. The results of the study are presented by a digital information system (DIS) model, for which the structure and priority digitalization technologies are determined. Along with this, classifiers of operating parameters have been developed, which, in combination with the conditions for the functioning of the DIS, provide a full-fledged application of trend and factor approaches to predicting heat supply to consumers. In terms of discussion, four areas of digitalization of HSO activities and their relationship with the achievement of the SDGs are identified. The key points that are introduced by the use of digital technologies as part of the DIS are identified. Taking into account the interdependence of the directions of HSO digitalization and the need for their joint implementation, problematic issues that must be taken into account when introducing digital technologies are characterized. The forecasting effects that can be obtained with the successful introduction of digital technologies in HSOs and further actions of the management of these organizations after the implementation of the DIS in the direction of complex digitalization of activities are determined.

N. G. Verstina
Moscow State University of Civil Engineering, Moscow, Russia

O. F. Tsuverkalova (✉)
Volgodonsk Engineering and Technical Institute, the Branch of National Research Nuclear University MEPhI, Volgodonsk, Russia
e-mail: oftsuverkalova@mephi.ru

N. A. Verstin
Peoples' Friendship University of Russia Named After Patrice Lumumba, Moscow, Russia

© The Author(s), under exclusive license to Springer Nature Switzerland AG 2023
A. Bencsik and A. Kulachinskaya (eds.), *Digital Transformation: What is the Company of Today?*, Lecture Notes in Networks and Systems 805,
https://doi.org/10.1007/978-3-031-46594-9_10

165

Keywords Digitalization · Heat networks · Forecasting · Heat supply organizations · Urbanized territories · Sustainable development goals

1 Introduction

The activities of many modern organizations are currently undergoing significant changes due to the emergence of digitalization tools, regarding which management will have to make decisions concerning the areas of use of digital technologies, the volume of their implementation in business management systems. Obviously, when making such decisions, they proceed from the expediency of incurring the costs of digitalization, combined with consideration of the benefits that will be received by the organization after its implementation. Under these conditions, the specifics of doing business of organizations and the features of creating products and services, which are the center of all digitalization issues, become extremely important.

From the standpoint of the listed aspects of digitalization, the most characteristic are heat supply organizations (hereinafter referred to as HSOs) that carry out their business in the urbanized territories of countries where district heating is predominantly used. The main activity of such organizations is the supply of consumers of heat energy by transferring it from a centralized source of generation to places of consumption through a system of heat networks (hereinafter–HN). This type of district heating is typical for the countries of the Eurasian continent, as well as North America, in each of which it is used at a different scale.

At professional discussion platforms, the question was repeatedly raised about the existence of a problem area that needs digitalization–the operation of heating networks laid from the source to consumers due to the presence of significant losses of heat energy during its transportation, which varies in different countries, but according to the information and analytical portal "EES EAEC. World Energy", is in the range from 5 to 15% [1]. At the same time, it is important to note that the vast majority of heating networks are laid underground, they have the most unfavorable operating conditions that violate their physical characteristics. It is well known that HSOs use certain methods of generating information on the operation of heat networks, but they are developed in each organization according to their own rules, therefore it becomes very problematic to compare data between organizations and at the same time there are still questions about the quality of primary information–subject to a long life cycle of heat networks (25 years and more on average) it is not always possible to make a full-fledged forecast of their condition.

At the same time, the United Nations Human Settlements Program (UN-Habitat), dedicated to the world's cities–"World Cities Report 2020: The Value of Sustainable Urbanization" notes that by 2030 it is planned to make the Global Sustainable Development Goals the goals of all people, businesses and governments of all countries of the world [2]. This is reflected in the multifaceted Decade of Action, that is also focused on sustainable urbanization based on the New Urban Agenda, which provides for "strengthening the role of affordable and sustainable housing and housing finance,

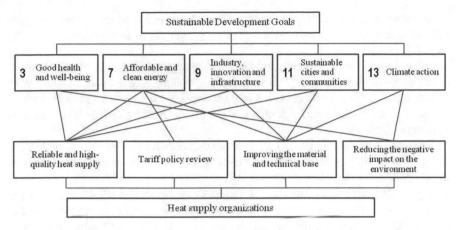

Fig. 1 Tasks for the HSO in the context of the SDGs

including the creation of a social habitat" [3]. Achieving these goals on the planned scale is possible only if it is possible to predict the situation with heat supply to consumers (Fig. 1), the beginning of which is the actualization of the approach to obtaining information at the level of individual HSOs using digitalization capabilities. The expansion of the scope of digital technologies will serve as the beginning of transformations in the organization itself in a number of areas, from improving the material and technical base to reducing the negative impact on the environment.

2 Materials and Methods

To date, countries using district heating have come to organize it in urban areas in different ways–from the creation of large combined heat and power plants that supply entire urban areas with heat energy, ending with boiler houses, the energy of which is designed for several compactly located houses. The Russian Federation, the countries of the "post-Soviet space" of Central Asia (Kazakhstan, Kyrgyzstan, Tajikistan, Turkmenistan, Uzbekistan) have a more extensive experience, less extensive, but nonetheless indicative of European countries–mainly Scandinavian (Sweden, Denmark). But at the same time, all countries have "one common denominator" of district heating systems–heat networks, which became the initial object of analysis in the study.

Researchers in different countries [4–7], as well as the state authorities of countries [8–13], were engaged in improving the efficiency of heat supply to consumers. At the same time, attention was paid to issues of both technical [14–16] and managerial nature [17–20]. However, so far there are quite a few works focused on the digital transformation of heat supply. In order to identify the most promising areas of digitalization, the authors conducted a preliminary analysis of the current state of the industry.

According to the information and analytical portal "EES EAEC. World Energy" in all the considered countries, there is a significant difference between the indicators "Total supply of heat energy" and "Final (energy) consumption", the difference in the values of which, along with own consumption, includes energy losses in heat networks (Table 1).

The data presented indicate that in modern conditions in the HSO, which are the primary source of information, there is still not enough information to assess losses, predict them, and most importantly, eliminate the causes of their occurrence. This provided the authors with the basis for the assumption, which was tested in the course of the study, that traditionally collected data on the operation of heat networks can be significantly improved through the use of digitalization opportunities, which will allow HSO management to operate with a large amount of data covering all significant factors of operation of heat networks.

In this regard, for the primary structuring of the source materials on the operation of heat networks, which can be used in forecasts, several provisions were used. Firstly, regardless of the country of origin, the constructive solution of heat networks is quite standard: a pipeline, insulating layers, in some cases, a channel in which they are laid. Significantly differ from the point of view of ground and underground, including channel and channelless laying, which over time is affected by negative factors of contact with the ground. Secondly, there are differences in operation between the linear part of heat networks and other, non-linear ones (compensators, gate valves, etc.), due to which the parameters of the heat energy flow are regulated. Linear parts are more subject to problems during operation. Thirdly, the types of damage to heat networks and failures in operation, which are associated with a certain type of gasket

Table 1 Data on the supply of heat energy and losses in networks

Country	General supply of heat energy, TJ	Losses in heat networks	
		TJ	%
France	169,194	13,140	7,8
Germany	457,481	40,772	8,9
Kazakhstan	329,930	32,227	9,8
Kyrgyzstan	12,424	941	7,6
Norway	24,005	2647	11,0
Russian Federation	5,113,878	397,451	7,8
Sweden	207,587	26,409	12,7
USA	369,846	55,477	15,0
Uzbekistan	133,833	7507	5,6

Source compiled by the authors according to [1]

and structural parts of engineering systems, have already been studied quite well
[21].

The analysis of the statistical data of heat networks in the Russian Federation,
carried out by the authors, showed that most of the problem areas in ensuring the reliability and energy efficiency of heat supply are associated with defects in pipelines.
The causes of defects in pipelines are shown in Fig. 2.

When assessing the risks of emergencies in heat networks, a fairly common
approach is related to the calculation of the probability of failure based on the
two-parameter Weibull distribution, the coefficients of which are determined on
the basis of many years of statistical observations for various types of laying and
insulation of pipelines, both for the entire set of defects and for their separate types.
Figure 3 presents graphs of the predicted failure probability obtained by the authors
for pipelines with a diameter of 301–600 mm, depending on the service life and type
of laying.

All these data are necessary, but not sufficient, to obtain reliable forecasts on the
scale of the entire heating network economy of the HSO. In particular, practical
experience shows that the use of this approach provides certain opportunities for

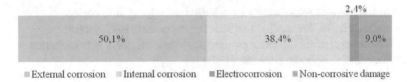

Fig. 2 Distribution of the causes of defects in pipelines that affect the energy efficiency of heat
supply systems

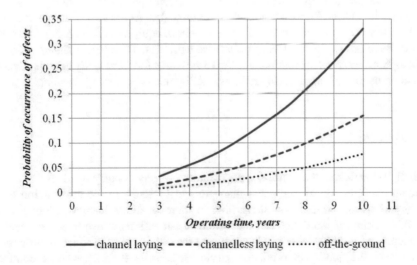

Fig. 3 The probability of occurrence of defects in pipeline sections

predicting the state of the HSO heat network economy, taking into account the age structure of heat networks, data on which were previously collected by the organization's staff. However, a certain "episodicity" in obtaining such data in modern practice should be replaced by a "total coverage" of operational parameters across the entire HSO due to the new opportunities provided by the use of digital technologies–the Internet of things, big data technologies, artificial intelligence methods allow process reengineering monitoring and diagnosing heating networks and making decisions at all stages of their life cycle.

The studies carried out earlier by the authors, which were used in this work as source material, showed that at present it is possible to predict the general nature of the trends in the behavior of heat networks during operation. At the same time, certain quantitative characteristics are observed, the degree of reliability and the possibility of using them in forecasts is determined by the quality of the information that was collected by the HSO management in the previous period of time. In particular, from the standpoint of applying the trend approach to forecasting, which assumes the identification of the regular component of the time series and the use of extrapolation of the aligned values of the predicted indicators, it has been established that in order to ensure the reliability of the forecast, it is necessary to ensure:

- the availability of reliable data for a sufficiently long time period;
- the uniformity of the applied methods for measuring parameters;
- the improvement of the used mathematical models and substantiation of their applicability in specific situations [22].

The application of a factorial approach to forecasting, based on the use of estimates of the degree of influence of various factors and their combination on the further development of the possible state of the control object, as a rule, uses models and methods designed to work with metric (quantitative) data, and in some cases, data presented in ranking (ordinal) and dichotomous scales. At the same time, the analysis of the causes of the occurrence and development of equipment defects in many cases cannot be reliably described using only these variables. Therefore, today the issue of developing new types of models that involve the use of fundamentally new approaches based on machine learning and artificial intelligence is becoming relevant.

3 Results

The digitalization of the sectors of the economy of any country involves the development and creation of integrated management systems on the information technology platform, which make it possible to increase the efficiency and quality of the functioning of industries and individual organizations, move to a new level of development, increase labor productivity and the efficiency of production processes, and ensure the quality of products (services). Despite the existence of common approaches in the field of informatization and digitalization, the solution of the problems of digital transformation of HSO is largely determined by their specifics.

The results of the research conducted by the authors made it possible to determine the main processes, the degree of digitalization of which is currently at a rather low level in most HSOs: these include, first of all, the processes of processing diagnostic information and assessing the state of structural elements of the vehicle based on the available data. Reserves for improving forecasting through the use of digital technologies have been identified. Firstly, information about the values of key operational parameters obtained from sensors in the course of current monitoring and control of the state of heating networks requires real-time processing in order to timely identify deviations in the course of technological processes and make operational decisions. Secondly, the processing of data received during periodic diagnostic examinations carried out by non-destructive testing methods, as a rule, is carried out by specialists on the basis of accumulated operational experience, which is a rather laborious process and does not exclude the possibility of errors associated with the influence of the human factor.

To use these reserves, the authors propose to develop a digital information system (hereinafter abbr.–DIS) for the integrated diagnostics of heat networks, which must be implemented in the HSO. The main elements of the proposed DIS in the projection of the life cycle of heat networks are shown in Fig. 4.

It is based on a large-scale database of operational parameters of heat networks, built using Big Data technologies based on interaction with other information and technical systems through a special data collection module, which determine new possibilities for predicting consumer heat supply situations. In the DIS, the database (DB) should collect both information on heat network projects and as-built documentation after their construction, as well as data on monitoring their technical condition (in the context of individual sections) at all stages of the life cycle of these engineering systems. The developed database structure is implemented on the basis of solving the problem of formalizing the subject area: the main objects in the sections of the heating network are identified, information about which should be stored in the database, as well as a list of characteristics that they may have. The entities that form the database are divided into three main categories:

- structural elements of heating networks,
- controlled technological parameters,
- defects in structural elements of heating networks.

For each category, a list of classifiers has been proposed that allows you to describe in detail each element of the entity, taking into account the possibilities of obtaining information in the conditions of using digital technologies. The list of main classifiers is presented in Fig. 5, however, as digital technologies are mastered in HSO, it can be significantly supplemented.

The collected data will be accumulated in the database and used by the analytical block of the system. The DIS analytical block consists of two modules. The first of them is used in the process of continuous monitoring of the operation of HN

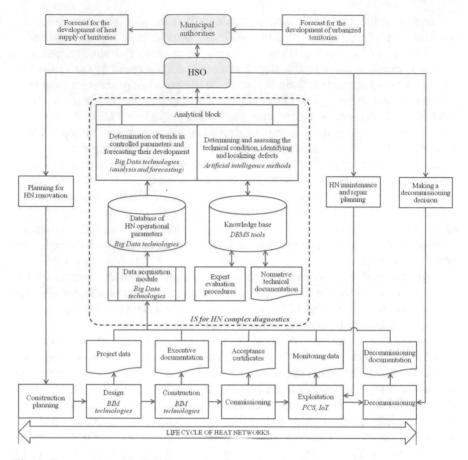

Fig. 4 The structure of the DIS for complex diagnostics of heat networks in the projection of the life cycle of heat networks

equipment and allows short-term forecasting of changes in the current values of operational parameters using modern packages of mathematical and statistical analysis and forecasting (regression analysis, time series analysis, analysis of "outliers", etc.). At the same time, it is possible both to predict trends in changes in individual parameters, and to build multifactorial models using lag variables that take into account the relationship between groups of indicators.

The second analytical module of the DIS is designed to assess the technical condition of heating networks based on the results of non-destructive testing in the process of diagnostics and predicting operational parameters in the long term using artificial intelligence methods (neural network modeling, machine learning, expert systems). Based on the processing results, a knowledge base is formed containing formalized descriptions (models) of the states of structural elements of heat networks at different stages of defect development. The formed knowledge base is necessarily

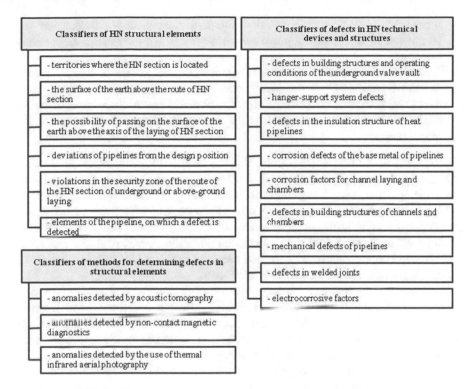

Fig. 5 Main DIS classifiers

supplemented with the data of normative and reference documentation and expert assessments of leading HSO specialists.

Studies have shown that when developing an analytical module in the practice of HSO activities, it is necessary to take into account the following conditions that ensure the full use of trend and factor approaches to forecasting for forecasting heat supply to consumers:

- selection of diagnostic parameters–their number should be sufficient to determine a specific defect and the factors of its occurrence;
- development of mathematical models that reflect the functional dependencies between the design characteristics of the HN element and diagnostic parameters;
- formation of a training sample–its representativeness significantly affects the quality of diagnostic and forecasting results;
- training of the diagnostic system for classifying the state of the object based on the involvement of experts to determine the quality of training and make the necessary adjustments.

Unlike the established practice, the use of digital technologies will provide:

– processing, identification of relationships and dependencies between the key parameters of technological processes and indicators of the state of engineering systems;
– formation of a system of analytical indicators that allows a comprehensive assessment of the state of the heating network;
– the use of an array of analytical information both in the current operating activities of the HSO and in the process of strategic planning for the development of the organization.

The presented DIS system, in combination with BIM technologies used at the design and construction stages, as well as using Internet of Things (IoT) technologies for collecting and transmitting data, will create a digital twin of the heating network, which will help to optimize equipment operation modes, to reduce coolant losses and heat energy, and, as result, to increase the efficiency of heat supply to consumers in urban areas in the context of the implementation of the Decade of Action, which is also focused on sustainable urbanization based on the New Urban Agenda in terms of creating a new habitat for people's livelihoods.

According to preliminary estimates, the use of the DIS will be able to provide a number of effects in the activities of HSOs, the source of which will be an increase in the accuracy of forecasts: in the field of organizing the activities of departments–improving their coordination when planning their work to provide consumers with heat energy, in the field of production processes–ensuring optimal modes of operation of engineering systems, in the field of finance–reduction of unproductive costs for the localization of emergency situations–by 10–15%.

4 Discussion

The obtained results of the study, concentrated on the DIS, allow in the future to raise the debatable issue more broadly–regarding the definition of the main directions of digitalization of HSO activities in the context of improving forecasting and their relationship with the achievement of sustainable development goals (Fig. 6). Let us give a brief description of the listed areas.

The first direction is the complex automation of the functioning of heat supply systems, by which the authors understand the use of a complex of automatic devices for controlling the production processes of heat supply, ensuring their complete remote control, regulating the operation parameters of heat networks, the operation of equipment and units of heat supply systems, accounting for the consumption of supplied and consumed resources. The introduction of software and hardware systems will allow solving the problematic issues of performing dispatching functions related to the optimization of heat supply modes, high electricity costs during its transportation, and others. The effect of the introduction of an automated control system for the functioning of heat networks can be up to 30% of all costs for production, transport and energy consumption [23].

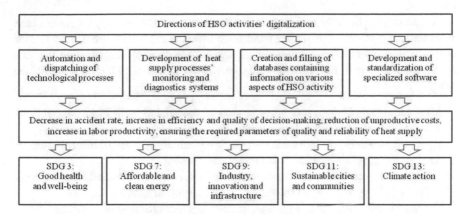

Fig. 6 Directions of digitalization of HSO activities

The second direction is the consistent implementation of the system of operational diagnostics and forecasting of the state of heating network elements into the practice of HSO activities based on real-time monitoring of their technological parameters (the starting option is provided for in the proposed DIS). Identification of deviations in this mode–directly in the course of technological processes will immediately eliminate their causes, without waiting for an emergency, which will significantly reduce losses associated with repair and restoration work. As a direct consequence of this, there is the elimination of interruptions in the provision of consumers with heat energy, an increase in costs and a decrease in the profits of HSOs, which significantly reduces the efficiency of the business of these organizations.

The third direction is the creation and filling of databases in various areas of HSO activity. First of all, we are talking about databases that allow accumulating data obtained in the course of monitoring the operational parameters of heating networks (the starting option is provided for in the proposed DIS). But at the same time, it is important to take into account the implementation of all other business processes in HSOs related to ensuring the functioning of heat supply systems (financial, personnel, information, etc.). The accumulation of such information of a complex nature will become the basis for the development of mathematical models that allow more accurate forecasting of HSO activities to provide consumers with heat energy, expanding the starting option–diagnostics of equipment and structural elements of the HN in the proposed DIS.

The fourth direction is the development and standardization of specialized software. Such software includes software products that allow simulating various heat supply processes, monitoring data processing programs, and others. Among the promising technologies, it should be noted the use of BigData technologies, as well as models based on artificial neural networks and machine learning, designed to form the basis of online and offline decision support systems. However, the development of electronic models of heat supply systems is faced with the lack of specialists in the HSO who own these software products.

As can be seen from the foregoing, all of the listed areas are closely interconnected and cannot be solved in isolation from each other. At the same time, one should also take into account the problematic aspects inherent in any organization that has embarked on the path of introducing digitalization. In technical terms, it should be noted:

- high requirements for computing resources associated with large amounts of information and real-time data processing;
- high requirements for data quality–insufficient amount of initial information, its low quality significantly affect the analytical results;
- the relative subjectivity of the forecasting results associated with the indirect influence of the thinking of the system developers and training experts;
- the need to organize reliable communication channels for data transmission using IoT technologies;
- ensuring information security of HSO information infrastructure facilities.

In organizational terms, the difficulties on the way of digitalization are associated with the high costs of developing and implementing the DIS, as well as with the need to develop new competency models and create a new corporate culture. Unfortunately, today in the HSO there is a certain shortage of employees with the required digital competencies, and therefore, most likely, a change in the organization's approaches in the field of personnel policy will be required.

Increasing the reliability of forecasting the activities of HSO as a whole by taking advantage of the digitalization of production and related other business processes of HSO in the context of the implementation of the above areas will open up new opportunities for the implementation of SDG 11, which refers to the quality of life of the population in urban areas, the number of which is UN forecasts will increase to 60% by 2030 [24]. The same projections note that urban areas also account for about 70% of global carbon emissions and more than 60% of resource use, to which heating systems make a significant "contribution". At the same time, it should be noted that, in accordance with SDG 7, it is important to take into account that energy is the dominant factor in the field of climate change, and it accounts for about 60% of total global greenhouse gas emissions (SDG 13). Accordingly, the digitalization of HSO activities will be able, in the context of SDG 9 priorities, "can unleash dynamic and competitive economic forces that create employment and income", provide "solutions to both economic and environmental problems, such as increasing resource efficiency and energy efficiency".

5 Conclusion

In general, digital transformation in order to obtain new opportunities for predicting heat supply to consumers will improve the reliability and quality of HSO services, reduce heat losses and duration of interruptions in heat supply, reduce time, material and financial costs for equipment repairs, optimize heat consumption at the level of end consumers. The first step in the digitalization of HSOs can be the creation of a DIS that takes into account the specifics of the provision of heat supply services to consumers and concentrates on the main production processes of these organizations. The next step in the direction of digital transformation should be the creation of a unified digital environment of the company, covering not only technological, but also corporate processes such as risk management, investment activities, production asset management, etc.

References

1. EES EAEC. World energy. https://www.ceseaec.org/. Accessed 23 Mar 22
2. A World Cities Report 2020: The value of sustainable urbanization. https://unhabitat.org/world-cities-report-2020-the-value-of-sustainable-urbanization. Accessed 23 Mar 20
3. New Urban Agenda. https://habitat3.org/wp-content/uploads/NUA-English.pdf. Accessed 12 Apr 23
4. Werner S (2017) International review of district heating and cooling. Energy 137, https://doi.org/10.1016/j.energy.2017.04.045
5. Lake A, Rezaie B, Beyerlein S (2017) Review of district heating and cooling systems for a sustainable future. Renew Sustain Energy Rev 67:417–425. https://doi.org/10.1016/j.rser.2016.09.061
6. Dylewski R, Adamczyk J (2012) Economic and ecological indicators for thermal insulating building investments. Energy Build 54:88–95. https://doi.org/10.1016/j.enbuild.2012.07.021
7. Liu G, Zhou X, Yan J, Yan G (2022) Dynamic integrated control for Chinese district heating system to balance the heat supply and heat demand. Sustain Cities Soc 88, art num 104286. https://doi.org/10.1016/j.scs.2022.104286
8. Electricity and heat annual questionnaire 2017–2021 and historical revisions. https://ec.europa.eu/eurostat/documents/38154/10015688/Ele-Instructions-2018.pdf/9f488d64-032a-46e2-a21a-6ea20afcdf92. Accessed 20 Mar 23
9. Heat Roadmap Europe 2050. https://www.euroheat.org/wp-content/uploads/2016/04/Heat-Roadmap-Europe-I-2012.pdf. Accessed 27 Mar 23
10. The energy efficiency strategy: The energy efficiency opportunity in the UK. https://assets.publishing.service.gov.uk/government/uploads/system/uploads/attachment_data/file/65602/6927-energy-efficiency-strategy-the-energy-efficiency.pdf. Accessed 18 Mar 23
11. The future of heating: Meeting the challenge. https://assets.publishing.service.gov.uk/government/uploads/system/uploads/attachment_data/file/190149/16_04-DECC-The_Future_of_Heating_Accessible-10.pdf. Accessed 15 Mar 23
12. Enhanced heating and cooling plans to quantify the impact of increased energy efficiency in EU member states. http://stratego-project.eu/wp-content/uploads/2014/09/STRATEGO-WP2-Executive-Summary-Main-Report.pdf. Accessed 27 Mar 23
13. On energy saving and on increasing energy efficiency and on amending certain legislative acts of the Russian Federation (2009) Federal Law N 261-FZ of 23 Nov 2009. http://www.consultant.ru/document/cons_doc_LAW_93978/. Accessed 03 Apr 23
14. Shan X, Wang P, Ren P, Zhao H (2016) The influence of central regulation methods upon annual heat loss in heating network. In: MATEC web of conferences, vol 54, art num 06004. https://doi.org/10.1051/matecconf/20165406004
15. Verstina NG, Evseev EG (2017) Technical-and-economic aspects of the analysis of the heating systems maintenance in the conditions of urban environment. In: MATEC web of conferences, vol 106, art num 08090. https://doi.org/10.1051/matecconf/201710608090

16. Gudmundsson O, Thorsen J (2022) Source-to-sink efficiency of blue and green district heating and hydrogen-based heat supply systems. In: Smart energy, vol 6, art num 100071. https://doi.org/10.1016/j.segy.2022.100071
17. Delangle A, Lambert R, Shah N, Acha S, Markides C (2017) Modelling and optimising the marginal expansion of an existing district heating network. Energy, № 140, p 209–223
18. Verstina N, Evseev E (2018) Reengineering of the management processes of information support of heat supplying organizations of heating systems of a city. In: MATEC web of conferences, vol 193, art num 05007
19. Verstina N, Badalova A, Evseev E (2019) Assurance of heating systems maintenance reliability through the creation of a risk management system of the heat-supplying organizations. E3S web of conferences, vol 97, art num 01019
20. Ketova K, Rusyak I (2021) Solving the problem of reliability improvement of the region heat supply system through optimal investment management. In: AIP conference proceedings, vol 2402, art num 070049. https://doi.org/10.1063/5.0071528
21. Verstina N, Evseev E, Tsuverkalova O, Kulachinskaya A (2022) The technical state of engineering systems as an important factor of heat supply organizations management in modern conditions. Energies 15(3):1015
22. Sysoev YS, Tsuverkalova OF (2019) The use of time series to improve the prediction reliability of the statistical characteristics of a controlled parameter. Meas Tech 61(10):973–978
23. Proshin AI, Shekhtman MB, Ladugin DV (2023) Integrated automation of heating networks as the most effective tool for increasing the efficiency of heat supply. https://www.krug2000.ru/pdf/kompleksnaya-avtomatizacija-teplovyh-setej-AVITE-4-2017.pdf. Accessed 20 Apr 23
24. Population Facts. https://www.un.org/development/desa/pd/sites/www.un.org.development.desa.pd/files/undes_pd_2020_popfacts_urbanization_policies.pdf. Accessed 12 Apr 23

RPA and Choosing Business Processes for Automation

Igor Nickolaevich Lyukevich, Artsrun Vrezhevich Melikyan,
and Inga Prokhorovna Sokolova

Abstract This study explores the procedure of Robotic Process Automation (RPA) process selection by investigating practical criteria for identifying suitable business processes for automation. Employing a mixed-method approach guided by critical realism philosophy and action research strategy, the research draws on historical developments of RPA and Process Mining (PM), uncovers their connection and synergy. Based on extensive desk research and real company data five main criteria were defined such as execution time, stability, process complexity, data type and failure rate. These criteria are crucial in evaluating processes for automation and contribute to a more systematic approach in RPA implementation. This way companies can be more accurate in their predictions of financial and operational return, and improve their decision-making model. The study underscores the significance of well-defined criteria in achieving successful RPA integration within various business processes. In conclusion, the aim is to build a foundation for improving business and decision-making efficiency as well as for further development in this field of knowledge.

Keywords Automation · Business process · PM · Process mining · Process selection · RPA · Robotic process automation

I. N. Lyukevich (✉)
Graduate School of Industrial Economics, Institute of Industrial Management, Economics and Trade, Peter the Great St. Petersburg Polytechnic University, Politechnicheskaya St., 29, St. Petersburg 195251, Russia
e-mail: lin.stu@yandex.ru

A. V. Melikyan · I. P. Sokolova
Graduate School of Management, Saint-Petersburg State University, Volkhovskiy St., 3, St. Petersburg 199004, Russia

1 Introduction

The modern world has several distinctive features. First of all, it is ever changing, and secondly, these changes are not just quantitative or qualitative, but they are very fast. The main reason for all of this is another key trait of our time which is IT-technologies. The Internet, different gadgets and software have permeated all spheres of our life, be it social, entertainment, business or anything else. The business sphere has experienced a very great change since receiving an email is incomparably faster than pigeon mail or messenger so most of the companies became both global and enormous in terms of their scale of operations. However, in order to be successful one should consider both effectiveness and efficiency, and for the purpose of solving these issues two new fields of expertise came into being: Process Mining (PM) and Robotic Process Automation (RPA).

Robotic process automation (RPA) is a rapidly emerging approach to automating processes using software robots that imitate human actions. To achieve this a process workflow is first recorded after which a virtual bot replicates the actions performed by humans in the graphical user interface of the application and automates their execution [1]. A virtual workforce consisting of multiple robots is created which enables the automation of various knowledge-related and back-office work that was previously carried out by human workers [2]. These virtual bots are integrated into existing software and can repeat tasks across multiple systems. Their configuration is guided by simple rules and business logic, and they can fulfill process steps independently of time, making them instantly scalable to handle increases in volume. As a result, RPA offers significant cost savings while also guaranteeing accuracy, consistency and ensuring first-time accuracy for tasks. Some of the potential use cases for RPA include data transfers and processing of high volumes of data.

1.1 History of RPA

The history of Robotic Process Automation (RPA) can be traced back to the early 2000s when the concept of screen scraping was first introduced. Screen scraping involved automating manual data entry tasks by extracting data from legacy systems and inputting it into new systems. However, this approach was limited in its scope and required significant manual effort to set up and maintain. The Fig. 1 graphically depicts the stages of development of this technology, which will be detailed below.

In the mid-2000s, a new generation of RPA tools emerged that were designed to automate tasks at the user interface level rather than relying on back-end integration with legacy systems. These tools were more flexible and could be implemented more quickly than previous screen scraping methods. They could automate a wide range of tasks, including data entry, data validation and report generation, and were often used to supplement manual processes.

Fig. 1 History development of RPA

The adoption of RPA began to accelerate in the late 2000s and early 2010s, driven by advances in technology and the increasing need for organizations to improve efficiency and reduce costs. New RPA tools were developed that could automate more complex tasks and integrate with a wider range of systems including cloud-based platforms.

Today, RPA is widely used in a variety of industries including banking, insurance, healthcare and manufacturing. It is estimated that the global RPA market will reach $13.74 billion by 2025, with many organizations investing in RPA to improve efficiency, reduce costs and enhance customer experiences. As the technology continues to evolve, it is likely that RPA will play an increasingly important role in the digital transformation of organizations across the globe.

1. In the 1990s, the foundation for RPA was laid with the automation of user interface (UI) testing. This involved testing the visual elements of interfaces to ensure their correctness and user-friendliness.

   At the time, there were fewer computer models available than there are today, but the appearance of PCs in the homes of ordinary people was increasing due to the popularity of the Windows 95 operating system. This led to a proliferation of UI testing requirements as screen sizes, and user needs became more diverse.

   As companies began to adopt agile development practices in the late 1990s and early 2000s, the importance of human capital and accelerating work processes became apparent. This resulted in the development of various scripts for automating UI testing and quality assurance (QA) to remain competitive.

2. In the 2000s, the first components of RPA that are still in use today emerged. This was made possible by screen scraping technology which involves automatically extracting data from applications for use in other contexts. This automation significantly increased the efficiency and effectiveness of many enterprises that needed to process large volumes of data. Banks and insurance companies, which aimed to streamline paperwork and improve compliance, were early adopters of this technology. These organizations also had the financial and human resources necessary to overcome the significant barriers to entry associated with automation at that time, such as the creation of complex IT environments, the search for highly skilled engineers and labor-intensive integration.

3. In the early 2010s, RPA experienced a major shift as it gained recognition from large enterprises. This was largely driven by a number of factors, including the

aftermath of the financial crisis which pushed businesses to reduce costs. Additionally, as companies began to recognize the urgent need for digital transformation, RPA emerged as a simple and affordable solution for many. As a result, RPA became increasingly popular and more and more companies started to rely on it to tackle their most important tasks.

4. Currently: Adoption of RPA by Small and Medium-sized Enterprises.

In the 2020s, RPA has become more accessible and is influencing all sectors of the economy. The proliferation of RPA technology is due to the decreasing cost of licenses which were once only available to large corporations. Vendors and system integrators are partnering to provide RPA-as-a-service solutions. Consequently, RPA is now available to businesses of all sizes and enables small and medium-sized enterprises to achieve high levels of performance.

Nowadays, small and medium-sized businesses are increasingly adopting RPA technology to streamline their operations. According to a 2021 Xerox report, 80% of small business leaders view task and process automation as vital for their company's survival in a highly competitive market. Furthermore, two-thirds of executives surveyed plan to upgrade their automation tools soon.

The changing demographics also play a significant role in RPA's popularity. Millennials and zoomers seek challenging and ambitious projects and find monotonous tasks uninteresting. RPA bots can automate many departments and roles in a company, freeing up employees' time for analytical and creative work. Effective implementation of RPA can improve productivity, reduce costs and enhance the overall performance of a business.

1.2 History of Process Mining

Process Mining is a bridge between Data Mining and Process Management. It is an approach to extracting, analyzing and optimizing processes based on data from event logs available in information systems. The Fig. 2 graphically depicts the stages of development of this technology which will be detailed below.

The first scientific theory, the purpose of which was the analysis and optimization of work processes, is "Scientific Management". At the turn of the 19–20th centuries, the theory of classical management was created by the efforts of the American researcher Frederick Taylor and his associates. It is based on the premise that there is a "best way" to do a particular job and the problem of low productivity can

Fig. 2 History development of process mining

be solved by using a method called "scientific timing". The essence of the method is to divide the work into a sequence of elementary operations that are timed and recorded with the participation of workers. As a result, this allows you to get accurate information about the time required to complete a particular job.

Process mining as a technique that involves the collection, analysis and interpretation of execution data extracted from event logs of Process-Aware Information Systems (PAIS) to understand a company's work routines, performance and social structure was described by van der Aalst at 2012. This method assumes that events are recorded sequentially and relate to specific activities that form cases. Transaction logs contain different types of information, including activity names, timestamps and user identifications which can be operationalized to achieve various objectives such as process understanding, control and performance improvement [3]. Process mining comprises three forms: discovery, conformance checking and enhancement. Discovery techniques aim to derive the control flow of a process without any prior knowledge while conformance checking detects structural deviations by comparing the as-is design with the to-be design of a process. Enhancement extends existing process models by utilizing timestamps to highlight bottlenecks, service levels, through put times, or execution frequencies. Process mining and RPA are complementary concepts, where process mining helps identify high-value automation candidates, quantify the economic value of corresponding initiatives, and eliminate costly and subjective manual process evaluations [4].

2 Methods

2.1 RPA Process Selection

Selecting the right processes to automate in a project is a crucial and challenging task as it directly impacts the success of the project Syed [5] tells that "A project can be deemed successful when the benefits achieved significantly outweigh the effort and resources invested by the organization." Hence, it is crucial for the organization to detect any recurring process and categorize it based on their relevance and significance to the organization.

According to Assatiani and Penttinen [6] in the context of RPA process selection, priority should be given to automating processes or tasks that are highly repetitive and have a fixed processing standard. But Wanner offers to look wider and provide six parameters [7]:

- Execution Frequency—Repetitive process tasks with a high volume of transactions, sub-tasks and frequent interactions between different systems or interfaces
- Execution Time—Average execution time of a process task
- Standardization—Streamlined process tasks with a-priori knowledge of possible events and outcomes of process task executions

- Stability—Process tasks with a low probability of exception and a high predictability of outcomes
- Failure Rate—Throwbacks ratio of process tasks, i.e. unusual and repetitive (partial) tasks until completion
- Automation Rate—Process tasks with a small number of steps that are already automated and offer less significant economic benefits.

Also, Mario Smeets in book Robotic Process Automation (RPA) in the Financial Sector [8] provide technical criteria for RPA:

- Degree of standardization—The more standardized, the sooner automation can take place
- Rule-based nature—The process must be fully rule-based in order to be fully automated. Otherwise, partial automation should be considered
- Process stability/maturity—The more stable a process is—i.e. the less frequently the process is adjusted—the more suitable automation is, since fewer adjustments of the bot have to be made in the operating procedure
- Complexity—The lower the complexity, the easier the automation
- Digitality of the data—Only digital data can be processed by the bot
- Structure of the data—RPA can only process structured data, i.e. data that the bot receives in the previously expected form. For example, arbitrarily formulated e-mails are not suitable; more advanced technologies are required for this
- Data type—Text and numbers are suitable, but pictures or handwritten data are less suitable. Supplementary technologies should be used for this purpose
- Applications involved—The more applications the process passes through and the higher the number of system breaks, the more sensible the automation with RPA.

To summarize: it is recommended that RPA initiatives require a thorough system of indicators to aid in the selection of process tasks, an evaluation methodology to determine the feasibility and benefits of automating process tasks, and an adaptable approach that considers the changing demands of process tasks.

RPA is one of the key areas of digital transformation. To assess the efficiency and effectiveness of new business processes, it is possible to use traditional analysis, [9] for example, "Canvas" by Osterwalder and Pigneur [10]. But in conditions of increased uncertainty, it is necessary to distinguish between "assessment" and "assessment of the consequences" of risks, [11] here comes to light the engineering and economic analysis, representing a combination of quantitative and qualitative methods, [12] algorithms of the fuzzy-multiple approach [13].

When robotizing business processes, it is necessary to assess the impact on the company's performance and calculate the operational value in combination with cost optimization–the possibility of reducing them to the optimal cost which will be achievable with simultaneous investment, production and financial planning [14]. In the work of Rodionov et al. seven information metrics are substantiated that allow us to assess the success of the transformation of the business process, among them ecology, social progress, economic development, innovation, brand, cooperation with the authorities [15]. Koroleva et al. have concluded that the support of the parent company creates a stronger competitive advantage for the introduction of new technologies, the combination of IT education and banking experience is a significant factor in the development of fintech companies [16]. In the study of Rudskaya et al. it is shown that the regional aspect has a significant impact on the effectiveness of the introduction of new technologies which is due to both extensive (due to the scale of resources) and intensive (due to productivity) reasons [17].

However, here we see the difficulty of combining two types of data (internal operational and data from the digital environment, some of which the company does not have direct access to). At the same time, significant amounts of information in the digital environment have an unstructured, textual format, and researchers suggest using natural language processing methods to encode feelings and emotions of comments and different posts in social networks, [18] using Python, Excel-the VBA programming language [19].

2.2   General Analysis and Best Practices

To keep up with the rapidly advancing field of Robotic Process Automation (RPA), it's important to have access to the most up-to-date knowledge about process selection approaches and criteria. In order to obtain this knowledge we search across various databases, including Google Scholar, Springer, Elsevier, ResearchGate, IEEE Xplore, AIS eLibrary and ACM Digital Library. Our search included specific terms related to RPA such as "Robotic Process Automation," "RPA," "Intelligent Process Automation," "Process Mining," and "Automation criteria."

Most authors proposed four to six stage approaches for the organization of RPA projects [1, 6, 20–22], which typically began with process identification and suitability assessment, based on process walk-throughs. However, despite the numerous papers available on the topic, the literature review highlighted a lack of generally valid selection criteria for RPA process selection. Some decisions were based solely on process characteristics [23], while others combined process suitability with minimum expected savings [24, 25]. Nonetheless, the information we were able to gather from the 12 selected papers will be invaluable in guiding future research and practical applications of RPA.

What's the most interesting are three papers [4, 7, 26], the concept of robotic process mining is introduced as an innovative approach to identify mature processes which makes use of interaction logs to identify mature processes with high automation potential. However, these papers have certain limitations such as the lack of formal characterizations for suitable automation routines and dependence on log data availability. Additionally, the approach is restricted to fully rule-based subprocesses. Despite some high-level approaches for prioritizing RPA processes, the literature review underlines the need for a more structured and systematic approach to identify, prioritize and select suitable RPA processes. This indicates the absence of a detailed and universal approach to address the challenges of selecting and implementing suitable RPA processes. As a result, further research is essential to develop a comprehensive and structured approach that can overcome these limitations and ensure the effective selection and implementation of RPA processes. For this purpose, there is a review of the most important papers in the fields of RPA and PM:

1. The article "Process Mining and Robotic Process Automation: A Perfect Match" presents a three-step approach for effectively combining process mining and RPA to improve process automation initiatives. The approach involves assessing RPA potential, developing the RPA application and safeguarding RPA benefits through continuous monitoring using process mining. The article also discusses criteria for selecting processes suitable for RPA, including scalability, repetitiveness, standardization, complexity and automation potential. The benefits of combining process mining and RPA are illustrated through a case study of the Purchase-to-Pay process. Overall, the article suggests that process mining and RPA are well-suited for achieving digital transformation by improving efficiency, speed, agility and compliance.

2. The article "Robotic Process Mining: Vision and Challenges" proposes the concept of Robotic Process Mining (RPM) as a class of techniques and tools to identify and assess candidate routines for automation using RPA bots. RPM tools take logs of user interactions with applications and generate executable specifications for streamlined and standardized variants of automatable routines. The article discusses the challenges and guidelines for realizing the RPM framework and proposes criteria for identifying which routines are suitable for automation using RPA tools. The criteria include routine tasks that involve structured data, have deterministic outcomes, involve user interactions, are repetitive and rule-based, and have potential financial benefits. The article emphasizes the importance of using a systematic approach, supported by RPM tools, to identify and prioritize candidate routines for automation using RPA tools based on a combination of automatically derived attributes and domain knowledge.

3. Paper "Importance of Process Flow and Logic Criteria for RPA Implementation" discusses the importance of process flow and logic criteria in recommending Robotic Process Automation (RPA) implementation. It presents a case study using a process discovery technique to identify a suitable candidate for RPA implementation, followed by examining the impact of automation on overall performance using simulation models. The methodology is well-structured and provides a clear understanding of RPA implementation. However, limitations of the methodology and the case study are not fully discussed and cost-effectiveness is not considered.

4. One of the important works is from the creator of the manifest of process mining and full professor at RWTH Aachen University, leading the Process and Data Science group and chief scientist at Celonis–Wil van der Aalst. In his article "Process mining and RPA: How to pick your automation battles?" he discusses the historical context of Robotic Process Automation (RPA) and process mining, and their interrelation. The paper shows how traditional Workflow Management

(WFM) and Business Process Management (BPM) systems have been limited to high-volume structured processes, and how RPA has made automation of routine processes economically viable. The paper also highlights that process mining helps identify process fragments that can be automated using RPA, and thus, process mining and RPA complement each other. The paper concludes by discussing the potential and challenges of RPA and how process mining plays a crucial role in addressing these challenges. The paper also discusses the changing frontier between tasks performed by humans and those performed by machines and algorithms, and how process mining can be placed in a larger context where work is distributed among machines and people.

3 Results

3.1 Research Design

See (Fig. 3).

- Philosophy and Approach to Theory Development

  Critical realism is a philosophical approach that recognizes the objective reality of the external world and acknowledges that our knowledge of the world is limited and can be improved through observation, experience and scientific investigation. Adopting this philosophy allows us to develop a deeper understanding of the underlying mechanisms and processes that drive RPA and its impact on business processes.

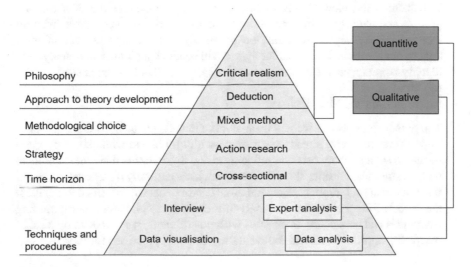

Fig. 3 Research description

In this study, we will be using a deductive approach to theory development. This approach involves developing a set of theoretical criteria based on existing literature and then testing them against empirical data. By testing our hypotheses against real-world data, we can refine and improve our theoretical framework.

- Methodological Choice

Our methodological choice for this study is a mixed-method approach, combining quantitative and qualitative techniques. This approach allows us to obtain a comprehensive understanding of the processes under investigation and to triangulate our findings to enhance the credibility and validity of our results.

- Strategy

We will be employing an action research strategy in this study. Action research involves a collaborative and iterative process of problem-solving, where researchers work closely with practitioners to identify, develop and implement solutions to real-world problems. This approach will allow us to work with practitioners to develop practical and applicable criteria for evaluating business processes for RPA.

- Time Horizon

The time horizon for this study is cross-sectional. Cross-sectional studies involve collecting data at a single point in time, providing a snapshot of the processes under investigation.

- Techniques and Procedures

To analyze our datasets, we will be using a combination of quantitative and qualitative techniques and procedures. We will be conducting interviews with practitioners to gain an understanding of their experiences with RPA and their perspectives on which processes are suitable for automation. We will also be using data visualization techniques such as process mining to visualize the flow of each process and identify inefficiencies, bottlenecks and compliance issues. We will then use expert analysis to evaluate the quantitative data, such as resource savings, against our theoretical criteria. Finally, we will conduct qualitative data analysis to identify common characteristics present in processes that have been successfully robotized.

- Research Data and Methodology

The several datasets that we will be analyzing include ten datasets corresponding to processes that have already been successfully robotized while showing good results in saving labor costs and reducing risks, five datasets that were robotized with no significant results, three datasets which are currently under evaluation but meets the criteria for robotization and three datasets that were deemed to not meet the criteria for robotization. By analyzing these datasets using process mining techniques and conducting interviews with practitioners, we will develop a set of practically applicable criteria for evaluating business processes for RPA.

3.2   Definition of a Pool of Theoretical Criteria

The implementation of RPA can vary based on the unique characteristics of each process or activity. Therefore, several criteria are used to identify suitable processes for RPA implementation. Below we conduct a summary of key criteria relevant to the management of business processes, based on existing literature. To define the pool of theoretical criteria that affect the success of robotization, a desk research approach was adopted. The research involved a comprehensive search of academic and industry-related literature, reports, and other relevant publications. The focus was on identifying the different factors that could impact the success of robotization based on previous studies and practical experiences.

Desk research is a method of research that involves collecting and analyzing existing data and information that is readily available from various sources. It involves searching for relevant literature, reports and other published materials that provide insights and information about the research topic.

The reason for choosing desk research for this study is because it is a cost-effective and efficient way of gathering information and insights about the research topic. Desk research is particularly useful when conducting a literature review or when trying to identify a pool of theoretical criteria that could impact the success of robotization.

Desk research is a research method that involves gathering information from existing sources, such as books, journals, websites and other literature. It is also known as secondary research, as the data collected has already been published or is available in some form. The aim of desk research is to analyze and synthesize existing knowledge on a particular topic, in order to inform further research or decision-making.

One of the main advantages of desk research is that it is a cost-effective and time-efficient way to gather information, as the data is already available and does not require primary data collection. It can also provide a broad overview of the existing literature on a particular topic which can help to identify gaps in knowledge or areas where further research is needed.

Desk research has a long history in academic research and has been used in various disciplines, including business, sociology, psychology and medicine. It is often used as a preliminary research method to gain a deeper understanding of a particular topic before conducting primary research.

Desk research can be done using various techniques, including keyword searches, citation analysis and content analysis. In this case, the researcher likely used a combination of these techniques to identify relevant literature on robotization and extract the relevant criteria.

One of the key challenges of desk research is ensuring the quality and reliability of the data collected. As the data is already published, it may be subject to biases or errors. To address this, the researcher must carefully select the sources used and critically evaluate the information presented.

Overall, desk research is a valuable research method that can provide a broad overview of existing knowledge on a particular topic. In the context of the chapter, it was a useful method for identifying a pool of theoretical criteria that can inform further research on the success factors of robotization.

Desk research, also known as secondary research, is a well-established method for gathering information from existing sources, such as academic literature, industry reports and online databases. As such, it is a widely used method across various academic disciplines and industries.

In terms of literature, there are many resources available on desk research, including textbooks, academic articles and online resources. Some examples of relevant literature on desk research include:

- "Research Methodology: Methods and Techniques" by Kothari [27]: This text-book provides an overview of various research methods, including desk research, and discusses the advantages and limitations of each approach.
- "The Literature Review: A Step-by-Step Guide for Students" by Ridley [28]: This book provides practical guidance on conducting a literature review, including how to identify relevant sources, how to critically evaluate sources, and how to synthesize information from multiple sources.
- "Market research in practice: a guide to the basics" by Hague et al. [29]: This book provides an overview of desk research methods specifically in the context of marketing and management research.
- "Secondary research: Information sources and methods" by Stewart and Kamins [30]: This book provides an introduction to desk research, including tips for identifying relevant sources and evaluating the quality of information.

After conducting a thorough desk research, a set of 5 theoretical criteria that could impact the success of robotization were identified. These criteria are as follows:

Overall, desk research proved to be a valuable method for identifying and defining a pool of theoretical criteria that could impact the success of robotization. The identified criteria will serve as the basis for the subsequent stages of the research which will involve empirical data collection and analysis to evaluate the relative importance of these criteria in the success of robotization (Table 1).

4   Discussion and Conclusion

Many RPA projects fail because automation turns out to be infeasible or they try to automate processes that are too infrequent or changing too fast. RPA needs to be approached more systematically using data-driven analyses methods. Selecting the right processes to automate is crucial for the success of a Robotic Process Automation (RPA) project. Prioritizing highly repetitive tasks with fixed processing standards is a common approach, but other parameters such as execution frequency, execution time, standardization, stability, failure rate and automation rate should also be considered (Fig. 4).

Additionally, technical criteria such as the degree of standardization, rule-based nature, process stability/maturity, complexity, digitality and structure of data, data

Table 1 Derived RPA criterias

Criteria	Comment	Paper
Execution Time (ET)	The longer it takes to complete a task, the more significant the potential benefits of automation	Davenport, T. H., & Kirby, J. (2016). Just how smart are smart machines? [31]
Stability (SB)	Adequate resources, including budget, staff and equipment, are required for successful implementation	Lacity, M. C., Solomon, S., Yan, A., & Willcocks, L. P. (2011). Business process outsourcing studies: a critical review and research directions. Journal of information technology, 26(4) [32]
Process Complexity (PC)	The technology infrastructure must be able to support the robotization process effectively	Willcocks, L., Lacity, M., & Craig, A. (2017). Robotic process automation: strategic transformation lever for global business services?. Journal of Information Technology Teaching Cases, 7(1) [33]
Data type (DT)	Text and numbers are suitable, but pictures or handwritten data are less suitable. Supplementary technologies should be used for this purpose	Leopold, H., van Der Aa, H., & Reijers, H. A. (2018). Identifying candidate tasks for robotic process automation in textual process descriptions. In Enterprise, Business-Process and Information Systems Modeling: 19th International Conference, BPMDS 2018, 23rd International Conference, EMMSAD 2018, Held at CAiSE 2018, Tallinn, Estonia, June 11–12, 2018, Proceedings 19 (pp. 67–81). Springer International Publishing [34]
Failure Rate (FR)	Throwbacks ratio of process tasks, i.e. unusual and repetitive (partial) tasks until completion	Chui, M., Manyika, J., & Miremadi, M. (2016). Where machines could replace humans-and where they can't (yet) [35]

type and number of applications involved should be taken into account. It is recommended that RPA initiatives require a comprehensive indicator system to aid in process task selection, an evaluation methodology to determine the feasibility and benefits of automating. As a result of our analysis using the desk research method with fundamental and established articles as the basis, we defined a set of five theoretical criteria which has a major impact on the success of robotic process automatization. These criteria are as follows: Execution Time (ET), Stability (SB), Process Complexity (PC), Data type (DT) and Failure Rate (FR).

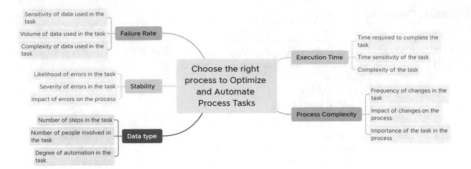

Fig. 4 RPA criteria MindMap

References

1. Huang F, Vasarhelyi MA (2019) Applying robotic process automation (RPA) in auditing: a framework. Int J Account Inf Syst 35:100433
2. Van der Aalst WM, Bichler M, Heinzl A (2018) Robotic process automation. Bus Inf Syst Eng 60:269–272
3. Li J, Wang HJ, Zhang Z, Leon Zhao J (2008) Relation-centric task identification for policy-based process mining. In: ICIS 2008 proceedings-twenty ninth international conference on information systems, pp 100
4. Geyer-Klingeberg J, Nakladal J, Baldauf F, Veit F (2018) Process mining and Robotic process automation: a perfect match. In: CEUR workshop proceedings, (CEUR-WS), pp 124–131
5. Syed R, Suriadi S, Adams M, Bandara W, Leemans SJJ, Ouyang C, ter Hofstede AHM, van de Weerd I, Wynn MT, Reijers HA (2020) Robotic process automation: contemporary themes and challenges. Comput Ind 115
6. Asatiani A, Penttinen E (2016) Turning robotic process automation into commercial success-case OpusCapita. J Inf Technol Teach Cases 6:67–74
7. Wanner J, Hofmann A, Fischer M, Imgrund F, Janiesch C, Geyer-Klingeberg J (2019) Process selection in RPA projects-towards a quantifiable method of decision making. In: 40th international conference on information systems, ICIS 2019, Association for Information Systems
8. Smeets M, Erhard R, Kaußler T (2021) Robotic process automation (RPA) in the Financial Sector. Springer Fachmedien Wiesbaden
9. Laidroo L, Koroleva E, Kliber A, Rupeika-Apoga R, Grigaliuniene Z (2021) Business models of FinTechs–difference in similarity? Electron Commer Res Appl 46:101034
10. Osterwalder A, Pigneur Y (2010) Business model generation: a handbook for visionaries, game changers, and challengers, vol 1. Wiley
11. Lyukevich I, Agranov A, Lvova N, Guzikova G (2020) Digital experience: how to find a tool for evaluating business economic risk. Int J Technol 11(6): 1244–1254
12. Kudryavtseva T, Skhvediani A, Brazovskaia V, Dracheva M (2022) Engineering economics: scientometric analysis of the subject area. Sustain Dev Eng Econ 3:5
13. Babkin A, Kvasha N, Demidenko D, Malevskaia-Malevich E, Voroshin E (2023) Methodology for economic analysis of highly uncertain innovative projects of improbability type. Risks 11(1)
14. Rodionov D, Koshelev E, Gayomey G, Ferraro O (2022) Model of global optimisation and planning of research and development costs of an industrial region. Sustain Dev Eng Econ 4:2
15. Rodionov D, Kryzhko D, Tenishev T, Uimanov V, Abdulmanova A, Kvikviniia A, Aksenov P, Solovyov M, Kolomenskii F, Konnikov E (2022) Methodology for assessing the digital image of an enterprise with its industry specifics. Algorithms 15(6)

16. Koroleva E, Laidroo L, Avarmaa M (2021) Performance of FinTechs: Are founder character-
    istics important? J East Eur Manag Stud 26:306–338
17. Rudskaya I, Kryzhko D, Shvediani A, Missler-Behr M (2022) Regional open innovation
    systems in a transition economy: a two-stage DEA model to estimate effectiveness. J Open
    Innov Technol Market Complex 8(1)
18. Konnikov E, Konnikova O, Rodionov D, Yuldasheva O (2021) Analyzing natural digital
    information in the context of market research. Information 12(10). (Switzerland)
19. Skotarenko O, Babkin A, Senetskaya L, Bespalova S (2019) Tools for digitalization of economic
    processes for supporting management decision-making in the region. IOP Conf Ser Earth
    Environ Sci 302(1):12147
20. Jimenez-Ramirez A, Reijers HA, Barba I, Del Valle C (2019) A method to improve the
    early stages of the robotic process automation lifecycle. In: Lecture notes in computer
    science (Including Subseries Lecture Notes in Artificial Intelligence and Lecture Notes in
    Bioinformatics). Springer, pp 446–461
21. Kokina J, Blanchette S (2019) Early evidence of digital labor in accounting: innovation with
    robotic process automation. Int J Account Inf Syst 35:100431
22. Santos F, Pereira R, Vasconcelos JB (2020) Toward robotic process automation implementation:
    an end-to-end perspective. Bus Process Manag J 26:405–420
23. Axmann B, Harmoko H (2022) Process & software selection for Robotic Process Automation
    (RPA). Tehnički glasnik 16(3):412–419
24. Willcocks L, Craig A, Lacity M (2015) Robotic process automation at Telefónica O2. Research
    on business services automation research objective. The Outsourcing Unit Working Research
    Paper Series 15/02, 28
25. Plattfaut R (2019) Robotic process automation-process optimization on steroids? In: 40th
    international conference on information systems, ICIS 2019, Association for Information
    Systems
26. Leno V, Polyvyanyy A, Dumas M, La Rosa M, Maggi FM (2021) Robotic process mining:
    vision and challenges. Bus Inf Syst Eng 63:301–314
27. Kothari CR (2004) Research methodology: methods and techniques. New Age International
28. Ridley DD (2012) The literature review: a step-by-step guide for students. In: Ridley D (ed)
    SAGE study skills, p 40
29. Hague PN, Hague N, Morgan CA (2004) Market research in practice: a guide to the basics.
    Kogan Page Publishers
30. Stewart DW, Kamins MA (1993) Secondary research: information sources and methods, vol
    4. Sage
31. Davenport TH, Kirby J (2016) Just how smart are smart machines? MIT Sloan Manag Rev 1:7
    Spring 2016
32. Lacity MC, Solomon S, Yan A, Willcocks LP (2011) Business process outsourcing studies: a
    critical review and research directions. J Inf Technol 26:221–258
33. Willcocks L, Lacity M, Craig A (2017) Robotic process automation: strategic transformation
    lever for global business services? J Inf Technol Teach Cases 7:17–28
34. Leopold H, van der Aa H, Reijers HA (2018) Identifying candidate tasks for robotic process
    automation in textual process descriptions. In: Lecture notes in business information processing.
    Springer, pp 67–81
35. Chui M, Manyika J, Miremadi M (2016) Where machines could replace humans-and where
    they can't (yet). McKinsey Q 2016:58–69

Predicting the Probability of Bankruptcy of Service Sector Enterprises Based on Ensemble Learning Methods

Dmitriy Rodionov⑩, Aleksandra Pospelova, Evgenii Konnikov⑩, and Darya Kryzhko⑩

Abstract This chapter focuses on developing an automated model for predicting the bankruptcy of trading enterprises. Due to the COVID-19 pandemic, many companies have suffered significant financial losses and continue to struggle with its negative consequences. This has resulted in an acute need for financial resources, making it crucial for credit organizations to enhance their credit scoring procedures. The chapter explores ensemble methods for predicting bankruptcy, including random forest, gradient boosting trees, and tree ensemble. By utilizing these methods, the researchers aim to improve the accuracy of bankruptcy predictions and provide credit organizations with a reliable tool for assessing the financial stability of trading enterprises. Given the current economic situation, the development of such an automated model has become more important than ever. By implementing these ensemble methods, credit organizations can make more informed decisions regarding lending and investment, which can have a significant impact on the stability of the financial market.

Keywords Digitalization · Ensemble learning methods · Bankruptcy · Credit scoring · Automatization

1 Introduction

Recently, predicting the probability of bankruptcy of companies has become especially relevant for credit institutions, investors, and the government [1]. The negative impact of the COVID-19 pandemic has resulted in many Russian companies in urgent need of borrowed funds, while the demand for their products has decreased due to the restrictions imposed. Assessment of bankruptcy risks enables to reduce the probability of default and increase profitability of investments. In this connection, more and

D. Rodionov · A. Pospelova · E. Konnikov · D. Kryzhko (✉)
Peter the Great St. Petersburg Polytechnic University, Polytechnicheskaya, 29, 195251 St. Petersburg, Russia
e-mail: darya.kryz@yandex.ru

A. Bencsik and A. Kulachinskaya (eds.), *Digital Transformation: What is the Company of Today?*, Lecture Notes in Networks and Systems 805,
https://doi.org/10.1007/978-3-031-46594-9_12

more attention is paid to credit scoring–a system of evaluating the creditworthiness of borrowers [2, 3].

Credit scoring is a technique to assess the likelihood of a borrower defaulting on a loan based on an analysis of his credit history and other factors. The scoring model helps lending institutions to make informed decisions about granting or denying credit. In other words, it is a process of assessing a borrower's creditworthiness by analyzing various factors, such as an organization's internal financial indicators, that can affect its ability to repay borrowed funds. Credit scoring is an important tool, because it can reduce the risk of non-payment of loans and increase the efficiency of investment [4–6].

One of the methods of predicting the probability of bankruptcy of enterprises is ensemble forecasting methods. These methods combine several forecasting models to improve the accuracy and reliability of the forecasts. Ensemble methods can include methods such as random forest, boosting, bagging, and others [7–9].

Ensemble methods are notable for their high accuracy, which is achieved by combining multiple models into a single system that produces a result based on a set of forecasts. They allow you to reduce the errors of individual models and thereby improve the quality of the forecast. Which is their main advantage.

Reducing the error and improving the accuracy of the forecast is achieved by using several forecasting models, which can take into account different aspects of the forecasted data and find more accurate dependencies between them. In addition, ensemble methods can account for some nuances, such as unbalanced data, outliers, and missing values.

The advantages of ensemble prediction methods are that they give more accurate results than individual models and are also robust to noise in the data. In addition, ensemble methods can account for different aspects of the data, which improves the quality of forecasting.

Predicting the probability of trade bankruptcy based on ensemble forecasting methods is a hot topic in today's world, as it can help lending institutions, investors, and the government make more informed decisions and reduce the risk of loss. In addition, it can improve resource efficiency and increase the productivity of the economy as a whole.

Thus, the use of ensemble forecasting techniques is an effective tool to improve the accuracy of forecasting the probability of corporate bankruptcy. This reduces risks for credit institutions, investors and the state and increases their profitability and stability.

2 Literature Review

The field of credit scoring is a multifaceted one for research. Researchers in many countries have contributed to the study of the factors influencing the bankruptcy of enterprises.

In [10] the author states that out of more than 150 models developed to date, even though they use more than 750 factors, such models are rarely implemented in practice. He explains this by the fact that the models are built according to the industry specifics and therefore may show lower efficiency for companies that are more influenced by factors not taken into account in the model. In this regard, the author develops an adaptive model based on factor selection, the application of an ensemble voting method and a genetic algorithm. The study was conducted on data from 912 Russian companies (456 bankrupt and 456 non-bankrupt) and the influence of 55 factors was evaluated. As a result, the accuracy of the method was 93.4%.

The author of [11] argues that ensemble methods are more accurate than other models. He develops his own model, which is based on quantitative estimation using Kohonen maps. The results show that such models lead to better predictions than those that can be achieved using traditional methods. The effectiveness of the model seems to be the most accurate due to the fact that it was evaluated using different data for different time periods.

In [12], the author shares the opinion of [11] that ensemble methods are more accurate, although the difference in performance with other methods is small. The method developed by the author is based on self organizing networks. As a result, it provides insight into financial models, which is relevant for most firms, as well as specific models that are relevant for non-standard companies, which, for example, have high liquidity exactly until they "collapse".

A description of various methods of predicting bankruptcy is described in [13]. The author gives a brief description of the method and finally gives a general review. He is convinced that machine learning is very effective in forecasting, but there is still a lot to strive for. First of all, it is necessary to diversify data sources, to use expert opinions, news and public reports in addition to reporting. It is also important to increase the functionality of predictive models to be able to determine which factor is more important in the company becoming bankrupt [14].

In [15] the author proposes the use of an alternative forecasting method–AdaBoost. He says that neural networks have an advantage in finding nonlinear relationships and are highly accurate in the presence of "noise". AdaBoost builds baseline classifiers sequentially using different sampling options. As a result, the new method reduces the generalization error by about 30% relative to the neural network error.

In [16] the author compares the three most commonly used statistical models–multiple discriminant analysis, linear probability model and logistic regression. Also considered how each model corresponds to the theory of financial crisis.

The author of [17] writes about the advantages of the logistic regression model for predicting bankruptcy. Using a sample of bankrupt and non-bankrupt firms, the author improves the logit-model using neural networks. The results show that this method predicts bankruptcy more accurately and offers promising results.

Work [18] is devoted to the analysis of the suitability of domestic and foreign models for predicting the probability of bankruptcy. The sample consisted of Russian enterprises of the manufacturing industry. As a result, the prediction accuracy of the constructed models is 84.7%. But at the same time, with the help of a binary classification tree, the model is able to establish the boundaries of normal values for the liquidity ratio and financial stability. Exceeding the limits may lead to the bankruptcy of the enterprise.

In [19] the author tries to overcome the imperfection of ensemble models, which is associated with their inefficiency for companies of a specific industry. The author develops his method, which is based on the biclustering method, which in turn looks for groups of firms, each of which is equally affected by the same factors, and on the ensemble method, which is necessary for the greatest retrospective of bankruptcy options. As a result, it is shown how a combination of these methods can improve the quality of forecasts.

A very important observation was made by the author of [20] that recently the diagnosis of bankruptcy of companies is extremely important both for business owners and for the state. He develops a model that is characterized by a fast and accurate approach using extreme gradient increase based on quadratic logistic losses. Three samples (Japanese, Korean and U.S. bankruptcy dataset) are used to test the quality of the model. According to the author, the results are superior to those of other machine learning methods for bankruptcy prediction.

A new approach to bankruptcy prediction was developed by the author of [21]. He used extreme gradient amplification and synthetic functions, a concept developed by the author for higher-level statistics, it is calculated as a combination of econometric indicators and arithmetic operations. The sample consists of data from Polish companies. The result is estimated in the prediction of bankruptcy from 1 to 5 year.

In [22] the author makes an important observation that the risk of company bankruptcy becomes the main reason for refusal of cooperation. The study was conducted on 1234 European companies.

The result of the experiment conducted by the author of [23] shows that the effectiveness of ensembles depends on the prevalence of positive values in the sample.

3 Methodology

The company's performance is described by a set of indicators that are reflected in its accounts. From the reporting data, performance evaluation indicators are calculated, which can be analyzed to predict the likelihood of bankruptcy using self-learning models.

At the first stage of the study, we obtained data from the accounts of 21 Russian bankrupt companies and 21 non-bankrupt companies. An important point in choosing the reporting period was the fact that even after the bankruptcy procedure, the company continues to exist in the form of external management, rehabilitation or bankruptcy proceedings and to function. This is because it is important for the company to use possible ways to repay its debts. Consequently, even after the bankruptcy procedure, the company may have its accounts published. However, if the company at the moment is already declared bankrupt, it cannot meet the requirements of creditors, and therefore the reporting data for the years after the declaration of bankruptcy is not suitable for the construction of the model. Based on the above aspects, it is necessary to use the statements of bankrupt companies for the period before the bankruptcy.

There are no time requirements imposed on the reporting of non-bankrupt companies.

By obtaining data from the companies' statements, indicators for predicting bankruptcy can be calculated, such as:

- Liquidity ratio;
- Financial leverage ratio;
- Net working capital ratio;
- Return on assets (ROA) based on profit before interest and taxes (PBIT);
- Asset turnover ratio;
- Equity ratio;
- Return on equity (ROE) based on profit before taxes (PBT);
- Current ratio;
- Working capital ratio;
- Proportion of fixed capital in total sources of funds;
- Interest coverage ratio;
- Return on sales (ROS);
- Sales-to-current-assets ratio;
- Current asset ratio;
- Return on investment (ROI) based on PBT;
- Return on equity (ROE) based on net income;
- Gross profit margin ratio;
- Asset utilization ratio;
- Working capital adequacy ratio;
- Management efficiency ratio;
- ROE based on PBT;
- Capital structure ratio (leverage);
- Profitability ratio;
- Efficiency ratio;
- Total asset turnover ratio.

These indicators were selected on the basis of existing bankruptcy prediction models. There are 7 domestic and 9 foreign models presented in Mazurov's textbook. Having analyzed each of the models, the indicators that can be calculated on the basis of Russian companies' statements were selected.

There are many methods of forecasting. However, to solve the problem of predicting bankruptcy, it is necessary to use models that can take as the resultant indicator a binary value: bankrupt–1, non-bankrupt–0. In other words, models that aim to determine the category of the object.

This chapter will use ensemble algorithms because they are supervised methods that combine predictions from two or more machine learning algorithms to produce more accurate results. Moreover, these algorithms explain implicit relationships between factors that, for example, regression cannot detect.

There are 3 main types of ensemble methods: stacking, begging, and boosting. Stacking uses different algorithms on the same data when training a model, such as regression. Begging trains, the algorithm several times on different samples, for example random forest. Boosting trains algorithms sequentially, each successive algorithm corrects errors of the previous one, i.e., each new sample includes data where the method was wrong in the previous sample, it is, for example, gradient boosted trees.

The algorithm for building the model looks as follows (see Fig. 1).

The program builds a model for predicting bankruptcy. Three methods of ensemble machine learning are implemented–Random Forest, Gradient Boosted Trees, Tree Ensemble. The sample is divided into training and test samples in the ratio of 70% to 30%.

1 method–Random Forest. Its basic unit is a decision tree, and the set of training methods, which returns classification or regression trees, is commonly referred to as CART (classification and regression trees). The method is based on a series of yes/ no questions about the input data. The main task of such questions, they are also called node separators, to determine which class the object belongs to. The use of CART implies composing such questions so that their answers lead to a reduction in the probability of misclassification (distribution) of the object, i.e., to a reduction in Gini Impurity.

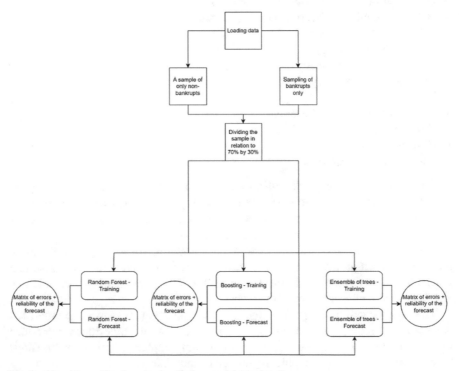

Fig. 1 Algorithm of bankruptcy prediction model construction

In KNIME for the Random Forest method it is possible to set the quality measurement criteria, according to which the separation will take place. These are Information gain ratio, Gain Ratio and Gini Index.

However, before defining each criterion, it is worth to form the notion of information entropy. Strictly speaking, it is a measure of sampling dispersion, i.e., it tells to what extent the number of objects of one class prevails over the other in the sample. A sample consisting entirely of bankrupts would have absolutely low entropy, equal to 0, and if in a sample bankrupts occupy some 50%, and non-bankrupts another 50%, then the entropy of such sample would be the highest (=1). Accordingly, the complexity of the prediction increases as the information entropy increases. The following is a formula by which we can calculate the information entropy:

$$ H = -\left(\sum Pi \cdot \log_2 Pi\right), $$

where Pi—is the share of class i in the dataset.

$$ P_i = \frac{D_j}{D}, $$

where D_j—the share of bankrupts/non-bankrupts in all the studied companies, D—the number of studied companies.

Information gain ratio–information gain ratio. The basic idea is that the lower the information entropy after dividing the sample with a yes/no question, the higher the information gain ratio. The main drawback of this criterion is that division divides the sample into subsets with minimal entropy, which results in the subset consisting of 1 object.

To exclude the above imperfection of the model, the Gain Ratio–gain coefficient was developed. It's intended to reduce deviation of Information Gain for strongly branched forecasts by introducing the normalizing indicator–inner information. The internal information formula is as follows:

$$ II = -(\sum \frac{|D_j|}{|D|} * \log_2 \frac{|D_j|}{|D|}) $$

For this sample Internal Information $= 0,25,678$.
    The gain is equal to:

$$ Gain\ Ratio = \frac{Information\ Gain}{Intrinsic\ Information} $$

Accordingly, the Gain Ratio for this sample $= 1/0,25,678 = 3,89$.
    The factor that gives the highest Gain Ratio is chosen as the question for the dividing node in the decision tree.

The third factor is the Gini Index. The initial purpose of this index is to determine the level of inequality in income and wealth of the population. But now it is adapted and used to optimize the division. It is calculated by the formula:

$$Gini = 1 - \left(\sum P_i^2\right)$$

For the study data, the Gini index $= 0.5$. Its application is similar to the Gain Ratio application.

2 method–Gradient Boosted Trees (GBT)–this method uses decision trees to create an ensemble of models. It works by building a sequence of models, each of which corrects the errors of the previous model. In GBT, each new tree is built based on the errors of the previous model. The GBT algorithm uses gradient descent (a method of finding the local minimum or maximum of a function by moving along the gradient) to optimize the loss function and reduce prediction error.

The 3rd method, Tree Ensemble, is a method that combines multiple decision trees into a single model. Each decision tree in the ensemble is constructed independently of the other trees and predicts a result, which is then combined into a single model. There are several ways to combine tree results, such as majority voting and mean.

The advantages of ensemble machine learning methods are that they can improve prediction accuracy and reduce overtraining [24, 25]. They can also handle large amounts of data and handle different types of variables. However, ensemble machine learning methods can be slow to learn and require more computing resources than single models. In addition, they may be less interpretable than single models because they combine multiple models into one.

4   Results

The bankruptcy prediction model includes the implementation of three methods–Random Forest, Gradient Boosted Trees and Tree Ensemble.

1 method–Random Forest. The Confusion Matrix for this method is as follows (see Table 1).

Two metrics–precision and recall–are used to analyze the quality of the obtained prediction. That is, accuracy and completeness, respectively (Table 2).

$$Precision = True\ Positives/(True\ Positives + False\ positives)$$

Table 1 Confusion Matrix for method 1

Actual/Predicted	1	0
1	9	1
0	0	4

Table 2 Statistics on the results of method 1

Attitudes towards bankruptcy/ indicators	TP	FP	TN	FN	Recall	Precision	F-measure	Accuracy
1	9	0	4	1	0,9	1	0,947	
0	4	1	9	0	1	0,8	0,889	
Overall								0,929

This is the part of the companies that the model categorized as bankrupt, even though these companies are really bankrupt.

$$Recall = True\ Positives/(True\ Positives + False\ negatives)$$

The proportion of bankrupt companies out of all bankrupt companies that the model was able to determine.

The best result was shown by the quality criteria–Information Gain Ratio and Gini Index, because they have the highest accuracy in predicting bankruptcy, which is the goal of the model (Table 3).

The 2nd method is Gradient Boosted Trees. The Confusion Matrix for this method is as follows (Table 4).

Total model error was 7.143%, model accuracy was 92.857%, precision of bankrupt companies forecast was 100%, non-bankrupt companies forecast was 80%, recall of bankrupt companies forecast was 90%, non-bankrupt companies forecast was 100%.

Model 3–Tree Ensemble. The Confusion Matrix for this method looks as follows (Table 5).

Table 3 Results of 1 method for each of the 3 quality criteria

Criterion/ Result	Error (%)	Accuracy (%)	Precision bankrupts (%)	Precision not bankrupt (%)	Recall bankrupts (%)	Recall not bankrupt (%)
Information Gain	14,286	85,714	90	75	90	75
Information Gain Ratio	7,143	92,857	100	80	90	100
Gini Index	7,143	92,857	100	80	90	100

Table 4 Confusion matrix for method 2

Actual/Predicted	1	0
1	9	1
0	0	4

Table 5 Confusion Matrix
for method 3

Actual/Predicted	1	0
1	9	1
0	1	3

Total model error was 14,286%, model accuracy was–85,714%, precision of bankrupt companies forecast was–90%, non-bankrupt companies forecast was–75%, recall of bankrupt companies forecast was–90%, non-bankrupt companies forecast was–75%.

Thus, the first and second models–Random Forest, Gradient Boosted Trees–are most applicable for the purpose of predicting the bankruptcy of the company. More clearly the results are presented in the table below (Table 6).

To understand the specifics of the model, it is necessary to establish common features in the companies that the model mistakenly assigned to the wrong category (hereinafter–the companies-errors).

Companies-errors of the model 1.

Company mistakenly represented as not bankrupt–TK A-MET LLC.

It has the highest return on assets based on earnings before interest and taxes, return on assets based on sales profit, return on equity and the highest profitability ratio of all bankrupt companies in the sample.

Misrepresented Companies Model 2.

The company mistakenly represented as not bankrupt is 4F LLC.

This company has a large share of net working capital in assets.

Error companies of model 3.

The company misrepresented as bankrupt is 82 REGION, LLC.

High return on equity based on pre-tax profit, return on equity based on pre-tax profit, and high total capital turnover ratio.

The company mistakenly represented as not bankrupt is TK A-MET LLC.

Table 6 Comparative matrix of model quality indicators

Criterion/ Result	Error (%)	Accuracy (%)	Precision bankrupts (%)	Precision not bankrupt (%)	Recall bankrupts (%)	Recall not bankrupt (%)
Random Forest	92,857	7,143	100	80	90	100
Gradient Boosted Trees	92,857	7,143	100	80	90	100
Tree Ensemble	85,714	14,286	90	75	90	75

Table 7 Alternative results of the three bankruptcy prediction methods

Criterion/ result	Error (%)	Accuracy (%)	Precision bankrupts (%)	Precision not bankrupt (%)	Recall bankrupts (%)	Recall not bankrupt (%)
Random Forest	81,818	18,182	84,4	75	90	64,3
Gradient Boosted Trees	79,545	20,455	83,9	69,2	86,7	64,3
Tree Ensemble	77,273	22,727	81,2	66,7	86,7	57,1

5 Alternative Results

As an experiment and to test the quality of the bankruptcy prediction model for Russian trading companies, the sample was increased from 42 to 174 companies, with 87 bankrupt and 87 non-bankrupts. The range of input data was increased by including Russian trading companies of the construction industry (Table 7).

As can be seen from the above table, all of the forecast quality criteria have decreased compared to the original ones. This is due to the fact that construction companies have their own unique specifics, which do not fully correlate with general trade companies. This experiment shs that bankruptcy forecasting models should be built with industry specifics in mind. And the more detailed this specificity is reflected in the influence factors, the higher the quality of the forecast.

6 Discussion

The results obtained largely confirm the theses presented in the literature review. Ensemble methods for predicting the probability of bankruptcy do show high accuracy and quality of the model, which agrees with the position of the authors of [11] and [12]. This suggests that ensemble methods can be an effective tool for predicting the probability of bankruptcy of Russian trade enterprises. However, it is also necessary to take into account the limitations associated with the application of ensemble methods, such as the need for a large amount of input data and high computing power requirements.

Despite the fact that ensemble forecasting methods have many advantages, such as improved accuracy and reliability, they also have their limitations:

- Complexity of interpretation: as a rule, the more models are used in an ensemble, the more difficult it is to interpret the prediction results. This is due to the fact that each model can account for different aspects of the data and make different decisions, which makes it difficult to explain the final results.

- Costly model storage: when a large number of models need to be trained, storing all those models can be a problem. This is especially true for large datasets, where each model can take up a significant amount of memory.
- Complexity of choosing the optimal ensemble: to obtain the best prediction, the optimal set of models and their weighting coefficients must be chosen. This can be a complex process, requiring experimentation and testing different combinations of models.
- The need for constant updates: since the data are constantly changing, the ensemble of models needs to be constantly updated and retrained to ensure high quality predictions.
- The need for large computational resources: Training and using an ensemble of models requires significant computational resources, which can be a problem for small companies or on personal computers.

In general, ensemble forecasting methods remain an effective tool for improving the quality of forecasting, but their use must be evaluated with specific conditions in mind. It is important to understand that these methods cannot guarantee accurate forecasts, but can significantly improve the quality of forecasting, especially in cases where the data contain noise.

Nevertheless, as the authors of [13] and [19] said, the quality and accuracy of the model could be improved by including external influencing factors in the sample. Thus, the result of the study would become more objective. This approach could be a recommendation for further research directions.

Moreover, in practice, the importance of taking into account industry specifics in the construction of bankruptcy prediction models of Russian companies was confirmed. As the authors of [10, 19] pointed out, in order to achieve a higher accuracy, it is necessary to build a model taking into account the factors that have the greatest impact on the industry under study. This chapter shows that adding construction industry companies to the sample consisting only of trading companies leads to the fact that the accuracy and reliability of the model falls.

Thus, the model for predicting the probability of bankruptcy of trade companies can be used in practice for lending institutions, investors or the state, for example, as one of the stages of credit scoring.

7 Conclusion

Thus, this chapter examined three ensemble methods for predicting bankruptcy of Russian trading enterprises: Random Forest, Gradient Boosted Trees and Tree Ensemble. The specifics of ensemble methods in general are described. It consists in the fact that these methods are able to find specific and non-linear relationships between objects. Random Forest and Gradient Boosted Trees showed the best results according to accuracy, completeness and reliability criteria. The reliability was 92.9%.

When the construction industry companies were added to the sample, the quality of the model fell, and the maximum reliability was 81.8% for the Random Forest method. This is due to the fact that each industry has a unique specificity, it is reflected in the influence factors. Therefore, in order to achieve the best result for the bankruptcy prediction model, it is necessary to select the appropriate factors for each industry.

Since models 1 and 2, Random Forest and Gradient Boosted Trees, have been the most effective, it is important to describe their specifics. According to the Random Forest model, all Russian companies are not bankrupt, have a high return on assets, capital and high profitability ratios. The Gradient Boosted Trees model classifies companies with a high percentage of working capital as non-bankrupt.

As a result of the research, we built a model for predicting bankruptcy of Russian trade enterprises, which has a high reliability of 93%. Moreover, this model has 100% accuracy in predicting bankruptcy, which is relevant for the purposes of understanding the financial well-being of companies.

The resulting model can be used by credit institutions, including banks, the state, investors. With its help it will be possible to achieve more balanced and objective financial decisions.

Acknowledgements The research is financed as part of the project "Development of a methodology for instrumental base formation for analysis and modeling of the spatial socio-economic development of systems based on internal reserves in the context of digitalization" (FSEG-2023-0008).

References

1. Kim H, Cho H, Ryu D (2022) Corporate bankruptcy prediction using machine learning methodologies with a focus on sequential data. Comput Econ 59(3):1231–1249
2. Rodionov D, Ivanova A, Konnikova O, Konnikov E (2022) Impact of COVID-19 on the Russian labor market: comparative analysis of the physical and informational spread of the coronavirus. Economies 10(6):136
3. Zhu Y, Xie C, Wang GJ, Yan XG (2017) Comparison of individual, ensemble and integrated ensemble machine learning methods to predict China's SME credit risk in supply chain finance. Neural Comput Appl 28:41–50
4. Rodionov DG et al (2022) Information environment quantifiers as investment analysis basis. Economies 10(10):232
5. Wang G, Hao J, Ma J, Jiang H (2011) A comparative assessment of ensemble learning for credit scoring. Expert Syst Appl 38(1):223–230
6. Ekinci A, Erdal Hİ (2017) Forecasting bank failure: Base learners, ensembles and hybrid ensembles. Comput Econ 49(4):677–686
7. Rodionov D et al (2022) Analyzing the systemic impact of information technology development dynamics on labor market transformation. Int J Technol 13(7):1548–1557
8. Mantoro T et al (2021) Neural information processing. In: 28th international conference, ICONIP 2021, Sanur, Bali, Indonesia, December 8–12, 2021, proceedings, Part III, vol 13110. Springer Nature

9. Parsotam P, Museba T (2021) A heterogenous online ensemble classifier for Bankruptcy prediction. In: 2021 3rd international multidisciplinary information technology and engineering conference (IMITEC). IEEE
10. Zelenkov Y, Fedorova E, Chekrizov D (2017) Two-step classification method based on genetic algorithm for bankruptcy forecasting. Expert Syst Appl 88:393–401
11. Du Jardin P (2018) Failure pattern-based ensembles applied to bankruptcy forecasting. Decis Support Syst 107:64-77
12. Du Jardin P (2021) Forecasting corporate failure using ensemble of self-organizing neural networks. Eur J Oper Res 288(3):869–885
13. Qu Y, Quan P, Lei M, Shi Y (2019) Review of bankruptcy prediction using machine learning and deep learning techniques. Proc Comput Sci 162:895–899
14. Rodionov D et al (2022) Methodology for assessing the digital image of an enterprise with its industry specifics. Algorithms 15(6):177
15. Alfaro E, García N, Gámez M, Elizondo D (2008) Bankruptcy forecasting: an empirical comparison of AdaBoost and neural networks. Decis Support Syst 45(1):110–122
16. Collins RA, Green RD (1982) Statistical methods for bankruptcy forecasting. J Econ Bus 34(4):349–354
17. Fletcher D, Goss E (1993) Forecasting with neural networks: an application using bankruptcy data. Inf Manag 24(3):159–167
18. Fedorova E, Gilenko E, Dovzhenko S (2013) Bankruptcy prediction for Russian companies: application of combined classifiers. Expert Syst Appl 40(18):7285–7293
19. Du Jardin P (2021) Forecasting bankruptcy using biclustering and neural network-based ensembles. Ann Oper Res 299(1–2):531–566
20. Le T, Vo B, Fujita H, Nguyen NT, Baik SW (2019) A fast and accurate approach for bankruptcy forecasting using squared logistics loss with GPU-based extreme gradient boosting. Inf Sci 494:294–310
21. Zięba M, Tomczak SK, Tomczak JM (2016) Ensemble boosted trees with synthetic features generation in application to bankruptcy prediction. Expert Syst Appl 58:93–101
22. Lukason O, Laitinen EK (2019) Firm failure processes and components of failure risk: an analysis of European bankrupt firms. J Bus Res 98:380–390
23. García V, Marques AI, Sánchez JS (2019) Exploring the synergetic effects of sample types on the performance of ensembles for credit risk and corporate bankruptcy prediction. Inf Fusion 47:88–101
24. Yeh JY, Chen CH (2020) A machine learning approach to predict the success of crowdfunding fintech project. J Enterp Inf Manag
25. Aljawazneh H, Mora AM, García-Sánchez P, Castillo-Valdivieso PA (2021) Comparing the performance of deep learning methods to predict companies' financial failure. IEEE Access 9:97010–97038

Intellectual Resources of Energy Sector Management in the Frames of Digital Economy

Natalia Ketoeva and Ekaterina Orlova

Abstract The purpose of this chapter is to examine the process of intellectual resources management, as well as to describe the knowledge management model in energy sector organizations. The relevance of the chapter is due to the rapid development of digital economy and organizations striving for innovative development within a competitive environment, which results in the necessity for efficient current human resources control mechanisms. The novelty is to develop a conceptual model of managing intellectual resources competencies in the organization of the energy sector. The specifics of this model involves a set of core competencies that define the effective implementation of a digital energy project. The tool used was the method of a knowledge management system construction, which implies identifying the points of knowledge implementation (core competencies), the connections between them, identifying the appropriate sources of knowledge (key experts) and intelligent managemental tools. The use of the resulting model by energy companies in the process of human resource management will enable the more efficient implementation and carrying out innovation projects, as well as holding a personnel policy of the company.

Keywords Digitalization of processes · Intellectual resources · Intelligent managemental · Knowledge management system · Intellectual resources management

1 Introduction

In the modern world, in terms of instability and volatility of markets, the successful operation of any enterprise is achieved not so much due to the material and financial resources necessary for production, but rather due to the work of highly qualified

N. Ketoeva (✉) · E. Orlova
Department of Management, NRU «Moscow Power Engineering Institute», 111250 Moscow, Russian Federation
e-mail: KetoyevaNL@mpei.ru

© The Author(s), under exclusive license to Springer Nature Switzerland AG 2023
A. Bencsik and A. Kulachinskaya (eds.), *Digital Transformation: What is the Company of Today?*, Lecture Notes in Networks and Systems 805,
https://doi.org/10.1007/978-3-031-46594-9_13

employees who are perceptive to various kinds of innovations [1, 2]. This is of particular value in the energy sector.

Science-based technologies of the energy industry are currently becoming the main driving force behind the economy development, both of a single country or a group of countries, and of the whole world. More widely appreciated are those specialists who possess not only technical knowledge, but also managemental skills, strategic thinking and economic knowledge. Gradually, new qualification requirements begin to form toward participants in the energy sector innovative activities, such as the ability for critical, analytical and logical thinking, quick responding to any situational changes, as well as the ability to constantly develop their creative skills. These qualities, more or less inherent in people, are called "key competencies" [3, 4].

All these circumstances result in the fact that it is impossible to manage creative intellectual workers using traditional methods. Consequently, there arises a need to create a universal model that allows one to effectively manage and optimally assign key specialists to innovative energy projects. This is being done to ensure the best correlation between their competencies and the projects they work on as well as a significant reduction in expenses [5, 6]. This fact determines the relevance of the research being under consideration.

The object of the study are the energy industry specialists' key competencies. The subject is the process of rational assignment of the organization's intellectual resources in terms of the transition to digital economy. The aim of the research is to examine the process of managing intellectual resources as well as to describe the knowledge management model in energy industry organizations.

The model of separate elements of the decision-making support system's interaction described by the authors enables achieving optimal assignment of key specialists in a multi-project environment. The essential feature of the created model is the possibility to assess the criterial characteristics of specialists, which possess a qualitative nature. The practical significance of the model is due to its versatility when planning and implementing projects in energy companies as well as the possibility to conduct an unprejudiced evaluation of specialists' key competencies while allocating project resources.

2 Materials and Methods

The now-established theory of intellectual capital, in one of its main concepts, approaches the division of a company's assets into material and non-material [7, 8].

The theory of managing intellectual-creative resources of an organization within the energy industry uses the main points of the human capital theory thus representing a person's creative abilities. These abilities, in their turn, are divided into a special type of his business qualities, proving their fundamental role in innovation activity. Key employees in innovative activity must possess such an exclusive set of qualities that includes an inclination for innovations, creative energy, operational and non-standard thinking [9, 10].

Considering the above points, human capital, from the perspective of its influence on the efficiency of activity improving, can be considered as a combination of three assets types:

- physical asset–physical qualities of employees, their health, etc.
- intellectual asset–knowledge, experience, skills, etc.
- creative asset–the ability to innovative and creative activity, regardless of its type.

In modern marketing conditions, the difficulty level in achieving competitive advantages for energy companies is increasing making it more complicated to solve any business problems–from occupying a leading position in the sales market to (competitive) survival.

The changing economic environment enforces companies to reconsider their approaches to the use of resources at disposal.

It becomes obvious, that in order to complete the task set, companies must have a certain set of resources, the combination of which forms the so-called key competencies and makes a particular company the owner of a unique competitive advantage.

Indeed, in terms of knowledge, information and technology spread, it is not enough to just have trained, "knowledgeable" specialists. The highest demand, especially in countries with developed economies, is for specialists owing certain characteristics:

- critical thinking ability;
- analytical and logical thinking;
- quick response to any situation;
- perceptivity to innovative activity;
- the need to constantly develop their creative skills.

The above-enumerated features have received a name "core competencies". Today this term has become central for the entire concept of managing intellectual and creative resources. Almost all scientists agree on competencies being person-ality traits. At the same time, it is necessary to distinguish exclusive and standard competencies [11, 12].

Exclusive personality traits are the exceptional abilities of a person, manifestating in obtaining unique results in those conditions that cannot be explained from the standpoints of traditional motivation theories.

These qualities include intellectual capital, innovative perceptivity, and creative energy. At the same time, one should not forget that such a creative person works together with professionals who, by all means, have high, but in a certain way, standard professional knowledge, skills, and abilities [13–15]. These characteristics have been named "standard competencies", in contrast to the first ones, that are called "key competencies".

The key competence of an energy company is such a competence, the presence of which allows it to solve problems that are beyond the strength of most other market participants; it sets a new standard for activities within the industry and thereby provides the owner with a competitive advantage.

The particular attention to this term is determined by the following facts:

Key competencies play a crucial role in the innovative nature of energy companies' activity.

The number of specialists with key competencies, as shown by modern research, is limited.

Key competencies cannot be directly measured since they manifest in a latent way, which requires a fundamentally new approach to assessing and managing them.

Every key specialist has its own set of key competencies, and the degree of their manifestation varies significantly.

Thus, to be able to compete successfully a high-tech energy organization needs to optimally assign projects to key specialists, which will ensure the highest possible correspondence of the competences to these projects and will significantly minimize costs.

The combination of high-tech organizations employees' key and standard competencies is exactly what forms the key competence of an energy company.

The methodology for building a knowledge management system consists of determining the points of knowledge use (key competencies) and the connections between them; identifying the suitable sources of knowledge (key specialists) as well as managemental tools. For each point identified, it is possible to describe the knowledge used (key competencies), for example, for predicting the quality of the project. The main steps of methodology are:

- knowledge sources determination (using the information about employees' competencies);
- constructing of the paths for the knowledge movement, which reflect the movement of knowledge from the sources to the points of their use;
- identifying the key employees (knowledge holders and consumers) in the constructed model;
- choosing the tools for knowledge exchange, accumulation, storage and distribution;
- developed model implementing and knowledge management process correcting.

Knowledge is in every company, and every company manages it. The difference lies in how conscious this management is. Knowledge can and should be managed, otherwise the company risks being squeezed out by new players who treat their assets in a more attentive way. Knowledge is as much an asset as finance, customer relationships, or a brand [16, 17].

The currently existing methods for assessing innovative qualities are based on generalized, integral indicators. For comparison, some of them might be pointed at.

Conceptual approaches to human resource management in modern business existing include:

- Skandia Navigator.
- Intellectual capital index (IC-Index).
- Technology Broker system.
- KPMI (Keys to Personal Mastery Inventory).
- Citation-Weighted Patents.

- Intangible Asset Monitor

The existing methods of assessing intellectual and creative resources, despite all their advantages, are characterized by some significant disadvantages:

- high subjectivity level;
- lack of reliable measurement metrics;
- they do not take into account the specifics of the energy companies' functioning.

Summing up all the above mentioned, we see that throughout the centuries-old history of labor research, human capital was most often assessed from the standpoint of an employee's level of qualification characteristics as a labor resource of an enterprise. In an industrial economy, when tangible costs of production were fundamental, such an assessment was logical and quite justified. Nowadays, in terms of the digital economy, this is getting clearly not enough. It is necessary to find a new method of assessing and managing intellectual capital.

The tools of multicriteria analysis in combination with the theories of fuzzy logic derived by Lofty Zadeh can serve as a basis for such a method development. Relying on the model of the decision-making process in fuzzy conditions, it can be claimed that the task of multicriteria analysis is reduced to ordering the elements of the set X using the criteria from the set G, allowing one to formalize fuzzy descriptions using fuzzy sets, linguistic variables and fuzzy relations.

So, if we represent a set of projects with their key competencies and a set of key specialists who need to be assigned projects to as efficiently as possible as the initial two sets, then these sets will be the input data, while the degree of candidates' compliance with the projects will be the output data.

3   Results

In connection with all the above mentioned, should describe a fuzzy model based on two binary fuzzy relations U and V. The first relation is built on two basic sets $X = \{\times 1, \ldots, xn\}$ and $Y = \{y1, \ldots, ym\}$, and the second one–on two basic sets $Y = \{y1, \ldots, ym\}$ and $Z = \{z1, \ldots, zl\}$. X here describes the set of projects being assigned to key specialists, Y–the set of characteristics (key competencies) of projects, and Z–the set of candidates for assigning projects to. In the focus of our interest, the fuzzy relation U meaningfully describes the profiles of energy projects, and V–the key specialists' profiles.

To determine their compliance with the projects, should use the compositions of the original fuzzy relations. So, (max–min)- and (max-prod)- compositions give information about the degree of candidate's compliance with the possibility of him being assigned a project to, and (min–max)- composition allows one to determine a specialist who is not suitable for a given project.

Thus, the maximax solution implies the maximum possible degree of compliance maximization. This method is very optimistic, namely, it does not take into account possible losses and, therefore, is the most risk-taking. The maximin solution is the minimum possible degree of compliance maximization. This method takes into account the negative aspects of various outcomes to a greater extent and is a more cautious approach to decision-making. The minimax solution is the maximum possible degree of compliance minimization. This is the most cautious approach to decision-making and considers all the possible risks most.

To be specific, should say:

X = {project N1, project N2, project N3, project N4, project N5};

Y = {internal motivation, non-standard thinking, systematic approach, teamwork, self-development};

Z = {Petrov, Ivanov, Sidorov, Vasiliev, Grigoriev}.

The specific values of the membership functions μU (xi, yj) and μV (yj, zk) of the considered fuzzy relations are reflected in Tables 1 and 2. Since the nature of linguistic variables (competencies) is exclusively qualitative, it is reasonable to take T = {NB, NM, Z, PM, PB}, where individual values are fuzzy sets with given membership functions as a term-set. The universal nature of this term-set allows to give each linguistic variable its inherent meaning. So NB stands for "very low", NM–"rather low", Z–"average", PM–"high", PB–"very high" value of the corresponding variable.

The practical task is to assign key specialists to individual projects so that to ensure maximum compliance of the key specialists' level of competencies with the required level of key competencies of projects.

Table 1 Fuzzy relation U of profiling projects

---	Internal motivation	Non-standard thinking	Systematic approach	Teamwork	Self-development
Project 1	PB	PB	PM	NM	NM
Project 2	PM	NM	PB	NB	NB
Project 3	NB	PB	Z	NM	PB
Project 4	NM	NM	NM	NM	NB
Project 5	Z	PM	PM	NB	Z

Table 2 Fuzzy relation V of key specialists' profiling

---	Petrov	Ivanov	Sidorov	Vasiliev	Grigoriev
Internal motivation	PB	PM	Z	PB	NB
Non-standard thinking	Z	NM	PM	NM	Z
Systematic approach	NM	NB	NB	PM	Z
Teamwork	NM	PB	NM	PM	NM
Self-development	PB	Z	NM	Z	NM

Since the considered relations satisfy all the requirements necessary for fulfilling their fuzzy composition, the result of the fuzzy composition operation can be displayed in the form of Table 3.

Table 3 analysis shows that based on the maximum values of the composition of the considered fuzzy relations membership function, the HR manager can be recommended to assign key specialists to the following projects: Petrov–projects 1 and 3; Ivanov–projects 1 and 4; Sidorov–project 3; Vasiliev–project 1; Grigoriev–projects 1, 3 and 4.

In terms of assigning the specialists in question: to project 1 it is most advisable to assign Petrov, Ivanov, Vasiliev; to project 2–Petrov, Ivanov, Vasilie; to project 3–Petrov and Sidorov; to project 4–Ivanov; to project 5–Sidorov and Vasiliev.

Should now consider a fuzzy (max-prod)-composition of the original fuzzy relations (Table 4).

Table 4 analysis shows that, proceeding from the same principles, the following projects can be assigned to these key specialists: to Petrov–project 3; to Ivanov–projects 1 and 4; to Sidorov–project 3; to Vasiliev–project 1; to Grigoriev–project 1.

Relying on the general principles of applied systematic analysis concerning the principle of multi-modelling, the following conclusion can be drawn: if the same results are obtained using different models, this may indicate the presence of a stable link or regularity between separate elements of the models. In respect to the considered fuzzy models, the coincidence of the results obtained on the basis of the (max–min)- and (max-prod)-compositions makes it possible to make more confident conclusions regarding the choice of certain specialists for assigning them to the corresponding projects.

Table 3 (Max–min)-composition of the original sets

---	Petrov	Ivanov	Sidorov	Vasiliev	Grigoriev
Project 1	PB	PB	PM	PB	PM
Project 2	PM	PM	Z	PM	Z
Project 3	PB	PM	PB	Z	PM
Project 4	PM	PB	PM	Z	PM
Project 5	Z	Z	PM	PM	Z

Table 4 (Max-prod)-composition of initial sets

---	Petrov	Ivanov	Sidorov	Vasiliev	Grigoriev
Project 1	PM	PB	Z	PM	Z
Project 2	Z	Z	NM	Z	Z
Project 3	PB	Z	PM	Z	Z
Project 4	Z	PB	Z	NM	Z
Project 5	Z	NM	Z	Z	NM

Table 5 (Min–max)-composition of initial sets

---	Petrov	Ivanov	Sidorov	Vasiliev	Grigoriev
Project 1	NM	NM	NM	Z	NM
Project 2	NM	NM	NM	NM	NM
Project 3	NB	NM	NM	Z	NB
Project 4	NM	NM	NM	NM	Z
Project 5	NB	NM	NB	NM	NM

Lastly, should analyze the results obtained in the course of the (min–max)-composition of the initial fuzzy relations (Table 5).

Since for fuzzy sets, indeed, the main values of the membership function are NB and PB, the worst case in terms of conclusions is the value of Z, which cannot give an informative answer to the question of the inconsistency of this or that candidate.

Thus, based on the three compositions analysis, we can conclude that it is necessary to supplement and specify the data being obtained with the use of (max–min)- and (max-prod)-compositions and. There is also some of its discrepancy with the data of (min–max)-composition of the two fuzzy sets display. For example, according to the results of the first two compositions, Sidorov is more corresponding with project 3, and according to the (min–max)-composition, the level of his correspondence is average. Similarly, with the same Vasiliev's and Grigoriev's compliance with project 1, the former one suits it more.

4 Conclusion

Based on the results presented, some conclusions can be drawn. Unlike the typical tasks of making individual decisions, within the described multicriteria problem of resource allocation, the decision-maker acts as an objective mediator. However, the manager can carefully intervene in the decision-making course, resolving local conflicts between possible assignments.

In a real situation, the multicriteria assignment problem's solution is unfeasible without the help of a decision support system. Its structure should include databases (the problem's description, information about projects and key specialists, criteria and their scales, key competencies of these sets' elements). The influence of objective and subjective factors on the procedures for finding a solution should also be taken into account. The developed interface allows a manager to work with the system with minimum preliminary training. It is important that the system allows one to analyze the problem step-by-step and develop his own preferences in the process of interactive work with the system.

The procedures of accelerated solution searching provide an opportunity to choose the solution's type. Step-by-step comparisons of the elements' characteristics (objects and subjects) allow one to make the most optimal assignments step by step.

Needless to say that these are precisely such systems–advisers and assistants–that help a leader to better form, justify and explain his policies to others, thus increasing the chances of making reasonable and far-sighted decisions.

References

1. Sobol O, Lobacheva A (2017) Technologies of human resources management for effective intellectual capital management in the organization. J Hum Resour Intellect Resour Manag Russia 30–34
2. Khan NU, LI S, Anwar M, Khattak MS (2021) Intellectual capital, financial resources, and green supply chain management as predictors of financial and environmental performance. Environ Sci Pollut Res
3. Glazov MM, Redkina TM, Mohammed H, Hussein H (2021) Intellectual resources of the organization as the basis of knowledge management. In: Science and business: ways of development, pp 104–106
4. Sokolov AA, Scherbakov MV, Tyukov AP, Skorobogatchenko DA (2018) Intellectual management decision support for the industrial energy resources providing systems. In: Scientific technologies, pp 30–37
5. Lyubavskya IN, Menzhulova S (2019) Problems of human capital and intellectual resources management. Chapteral problems of agricultural science, production and education. In: Materials of the V international scientific and practical conference of young scientists and specialists (in foreign languages), pp 192–195
6. Reznikova KV, Gaponenko TV (2021) Modern directions of management of intellectual resources of the organization. Development of economy and management under conditions of transition to digital economy. Collection of scientific papers. Under the general editorship of K.A. Barmouths. Rostov-on-Don, pp 267–270
7. Boyko AE, Shushunova TN (2019) Optimizing management of intellectual resources of small innovative enterprises. J Successes Chem Chem Technol 12–13
8. Shkuratova MV, Gaponenko TV (2020) Intellectual resources as an object of management. Economy and management: current issues, achievements and innovations. Collection of chapters based on the materials of the international scientific-practical conference of young scientists. Ministry of Science and Higher Education of the Russian Federation, Don State Technical University, pp 272–275
9. Yakovleva E, Bulletin V (2018) Organizational levels of management of intellectual resources of workers at the enterprises of the industry. The faculty of management SPBGEU, pp 206–212
10. Yakovleva EA (2018) Management of intellectual resources of employees in the conditions of innovative development in the digital economy. Creat Econ Mag 1073–1088
11. Lvov VM, Golubeva NY (2017) The problem of management of intellectual resources of organizations. J Hum Factor Probl Psychol Ergon 55–59
12. Nikishina AL (2017) Audit of the intellectual capital as the basis of effective management of labor knowledge and resources. Karelian Sci J 122–125
13. Ustinova Liliya N (2017) Digital technology in the management of intellectual resources and innovative activity of the enterprises. In: Digital transformation of economy and industry: problems and prospects, pp 509–532
14. Galchenko SA, Antropov DV, Komarov SI, Zhdanova RV, Kirillov RA (2020) Innovative approaches to the formation of an intellectual system of support of decision making during the solution of tasks of management of land resources. In: Iop conference series: earth and environmental science
15. Melnikov ON, Yaremchuk AP (2018) Management of intellectual and creative resources of the organization, providing support and growth of competitiveness of innovation-active enterprises. J Creat Econ 1257–1272

16. Aron O'Cass, Phyra Sok (2013) The role of intellectual resources, product innovation capability, reputational resources and marketing capability combinations in firm growth. Int Small Bus J 32(8):996–1018
17. Tietze F, Vimalnath P, Aristodemou L, Molloy J (2023) Crisis-critical intellectual property: findings from the COVID-19 pandemic. In: General economics. https://arxiv.org/ftp/arxiv/pap ers/2004/2004.03715. Accessed 21 Aug 2023

Printed in the United States
by Baker & Taylor Publisher Services